Next Generation Phone Systems

How to Choose and Implement PBX Alternatives

By Dave Krupinski,
Charlie Schick, Ph.D.
and
Brian McConnell

CMP books

NEW YORK & SAN FRANCISCO

Next Generation Phone Systems

Published by CMP Books
An imprint of CMP Media Inc.
Converging Communications Group
12 West 21 Street
New York, NY 10010

ISBN 1-57820-038-5

For individual orders, and for information on special discounts for quantity orders, please contact:

CMP Books
6600 Silacci Way
Gilroy, CA 95020
Tel: 800- or 408-848-3854
Fax: 408-848-5784
Email: cmp@rushorder.com
Web: www.cmpbooks.com

Distributed to the book trade in the U.S. and Canada by
Publishers Group West
1700 Fourth St., Berkeley, CA 94710

Manufactured in the United States of America

Contents

Chapter 9
IP-PBX Market Overview

Chapter 10
The Virtual Phone System

Chapter 11
Virtual Phone System Market Overview

Chapter 12
The Next Generation Telephone and Terminal Device

Chapter 13
Next Generation Approaches of Traditional PBX Manufacturers

Chapter 14
A Look Into the Future

Preface

The way we communicate is changing. In the past few years we've seen the introduction of a variety of new mediums for both business and personal communication. E-mail, the Web, fax, Internet chat, videoconferencing, and whiteboarding are just a few examples of technologies that are playing an increasingly important role in the way we communicate every day.

In fact, many analysts believe that these newer forms of communication are even more popular today than voice communication via the telephone. While this point may be arguable, no one disputes that fact that the use of these new mediums is growing at a much faster rate than voice communication.

If this trend continues—and there is every indication that it will—in just a few years, voice communication by telephone may actually account for a small minority of overall business and personal communication. Although voice communication is no less important than any other form of communication, it is undoubtedly becoming less popular than other forms of communication.

Just as the way we communicate is changing, so too are the tools that we use for communication. The PBX (private branch exchange) is a core component of the business communication infrastructure. However, the traditional PBX is a tool that has been optimized specifically for voice communication. In a world where the way we are communicating is changing, so too will the traditional PBX.

Already, we're seeing the introduction of a wide variety of communication solutions designed to do much more than provide voice communication. The "Next Generation" solutions that are arriving on the market today are designed to take advantage of the converging worlds of computers and networking to support data, voice and video.

Today, vendors are crowding the marketplace with a broad spectrum of solutions based on a variety of architectures. Anyone looking at this market for the first time will find it challenging to understand and evaluate these solutions before making a buying decision.

So what will your next phone system look like? Not long ago, answering this question was much easier than it is today. When buying a phone system meant selecting a proprietary black box, customers were often more concerned with the name on the box, than the technology inside. The technology was unapproachable and proprietary. As long as the system worked reliably and provided a standard feature set, customers weren't expected to care how it got the job done.

Today, however, anyone considering the purchase of a new phone system is required to understand and compare a variety of technologies, feature sets, platforms, standards, and strategies before making a selection. Not only do you need to consider the offerings of established manufacturers—Lucent, Nortel, Siemens, etc.—but you also need to evaluate the latest and greatest products from enterprising, young startups.

If you find yourself in the process of selecting a new phone system, then this book is for you. In this book, we'll explore a whole new generation of communication solutions and provide you with practical information and advice to help you make sense of this exploding market and select the appropriate solution for your business.

We'd like to thank the following companies for their valuable insights, assistance, and contributions to this book:

3Com Corporation, Alcatel, Artisoft, Inc., AltiGen Communications, Inc., C-Phone Corporation, Cisco Systems, Inc., Circa Communications, Ltd., COM2001.com, Comdial Corporation, dba Telecom Corporation, Dialogic Corporation, The Enterprise Computer Telephony Forum, Ericsson, Hello Direct, Inc., Interactive Intelligence, Inc., InVADE International, Jetstream Communications, Linx Communications, Inc., Lucent Technologies, Mantis, S.A., Microsoft Corporation, Mitel Corporation, Motion Media Technology, Ltd., Netergy Networks, Inc., Nortel Networks Corporation, Pingtel Corporation, Polycom, Inc., Praxon, Inc., RADVision, Siemens, Shoreline Communications, Inc., ThinkLink, Inc., Vertical Networks, Inc., VirtualPBX.com, Inc., Webley Systems, Inc., Wildfire Communications, Inc.

Dave Krupinski

dave@trekmail.com

Charlie Schick, Ph.D.

charlie@trekmail.com

Brian McConnell

brian@trekmail.com

FOREWORD

By Zigurd Mednieks

Why read this book? Indeed, why consider buying a non-traditional telephone system? The old answers to these questions have not served us well. IP telephone systems, and other packet-based alternatives to traditional systems that connect voice circuits were built with the expectation that the inherent technical superiority of TCP/IP networks over switched-circuit networks would cause traditional products to be displaced. The reality was people didn't care, as long as it worked. Then, as computer networks became larger and more valuable to most businesses than their voice telephone system infrastructure, the lower costs of wiring, operating, and maintaining packet telephony systems were supposed to carry the day, but did not. Bypassing expensive international long distance charges proved to be a niche market. Today, more than 95% of telephone systems sold use an architecture that is decades old.

In fact, some current telephone technology dates back to the very first commercial systems, about a century ago. Buy an antique telephone and plug it in (you might have to buy a plug adapter) and it will probably work.

If the engineering charms of packet switching, or the lure of looking at your telephone traffic with the same tools you use to make sure your data networks are ship-shape, or if the need to minimize your costs to call Kazakstan are all less than compelling, why should you consider telephony servers and packet telephony? The new, more convincing reasons have their source in the Internet, and e-commerce in particular.

The reason you should take a perfectly functional but also conventional telephone system and chuck it (or, more likely, sell it on Ebay) is that if every business must become an e-business using e-commerce to survive, then every component of a business, including the telephone system, must be selected with Internet e-commerce in mind. And this is where telephony servers and packet telephony shine. Distributed object technology on Windows 2000, Unix, and Java platforms enable you to knit together Web and telephone interaction. A customer that buys a product online should be able to track his purchase over the phone. A customer viewing a Web site should be able to interact with a sales agent via a continuum of technology, from online chat to Internet phone to plain old telephone, and the context of his interaction should be preserved across any shift in the medium of interaction. In short: buy your phone system based the premise that it will be part of your Web site.

In the age of Internet commerce telephony still matters, and it matters very much. Ten times as many cellular handsets as PCs will be manufactured this year, and a hundred times as many as handheld PDAs. The cellular network build-out rivals, and in places surpasses, the spread of the Internet in terms of cultural and economic significance. It would be short-sighted and lacking in global perspective to limit the implementation of e-commerce to customers who prefer to use computers and Web-browsers.

Computer makers see the overwhelming numbers associated with wireless mobile telephony and covet them. They are scheming to add personal computing and Internet access to the mobile experience. The well-equipped customer of the future will have a personal area network (PAN) linking a communications device, a computing device, and one or more input devices like a headset, camera, pen etc. Size limitations and other ergonomic factors mean that every possible form of human interface, certainly voice, will have to be enlisted in interacting with Internet-based servers. This future, driven by commerce, is a lot more likely than the earlier justifications for communications convergence based on the desirability of packet technology, tidying up network management, or long distance pricing.

The reason to take that first step now, even if you do not have the plans and budget in place to tie your telephone system into the rest of your e-commerce infrastructure is that failing to start promptly is the leading reason for failing to complete a project on time. You will be called upon to make your products and services available to the mobile Internet-connected customer, and soon. The expectation of convergence, no matter its exact form, is already firmly embedded in the minds of everyone that has both an Internet connection and a mobile phone. It is a prediction so safe that it takes no clairvoyance to say that businesses that are the most available and convenient to customers will prosper more than others.

Dave Krupinski is better prepared than anyone I can think of to guide you on this journey. Telephone systems, even the new kind, are long-lived. A telephony server will weave itself into the fabric of your business as much as your key enterprise software systems. Mistakes are costly. This book enables you to gain all the benefits of Internet-connectable telephone systems while avoiding any pitfalls. Those pitfalls, by the way, are fewer and farther between than they were in the first blush of packet telephony hype. The systems and their makers are mature and reliable. They are increasingly easy to deploy, use, and manage. If any doubts remain in your mind, this book will carry you across that gulf of skepticism and prompt you to action. Be ready.

Zigurd Mednieks is a consultant, entrepreneur, and author. He knows more than is good for you about unified messaging, computer telephony, IP telephony, voice over DSL, wireless telephony, and Windows-based multi-tier Web-oriented programming. Zigurd makes a living advising people how to avoid wrack and ruin in their product development projects. He designs systems, evaluates technologies, writes code, and solves problems. He also knows a lot of people in the business, so problems that can't be solved by brains, he can solve with money and influence. Zigurd lives in Harvard, Massachusetts, his e-mail address is Zigurd_Mednieks@msn.com.

Chapter 1

The Evolving Phone System

1.1 Introduction

In this chapter we'll explore the features and architecture of a traditional PBX—the type of phone system that most businesses have installed in their office today. We'll see that the traditional PBX is based on a proprietary design, optimized specifically for voice communication. Attempts to integrate the traditional PBX with computer and data networks have achieved limited success.

The revolution brought about by the computer, and more recently the Internet, has given rise to entirely new ways of communicating—and an entirely new generation of communication products. This chapter ends with a discussion of the factors that have motivated and influenced the development of the next generation phone systems on the market today.

1.2 What is a PBX?

A private branch exchange (PBX) is an on-premise facility that interconnects internal telephones and provides access to the Public Switched Telephone Network (PSTN). Initially, PBXs were smaller versions of the same equipment that was installed in the phone company's central office (CO) sites. Accordingly, they were viewed as *branches* of the CO installed in *private* locations. The term *exchange* refers to the PBX's primary task of switching voice from an input device to an output device.

Originally, these systems required manual switchboard operators to make connections between input and output devices. In the 1920s, automatic systems began to replace the manual operators. These first automatic systems were called private *automatic* branch exchanges (PABX) since they did not require an attendant to perform the switching. These systems used electromagnetic technology and analog signaling.

In the 1970s a new generation of digital PBXs was introduced that used electronic technology and digital switching. The PBX of today—although improved and enhanced since the 1970s—is based on the digital architecture. Throughout this book, the digital PBX of today is referred to as the *traditional PBX*.

The primary task of a PBX is to create a voice circuit between a user of the PBX and the PSTN. Additionally, a PBX can facilitate internal voice communication by creating a voice circuit between PBX users. A PBX provides a range of basic and advanced services. The basic services of a PBX include:

- Allowing multiple users to share a limited number of connections to the PSTN to place and receive telephone calls.
- Allowing users to call one another without using the PSTN.
- Allowing users to perform simple call control activities, such as transferring calls and placing calls on hold.

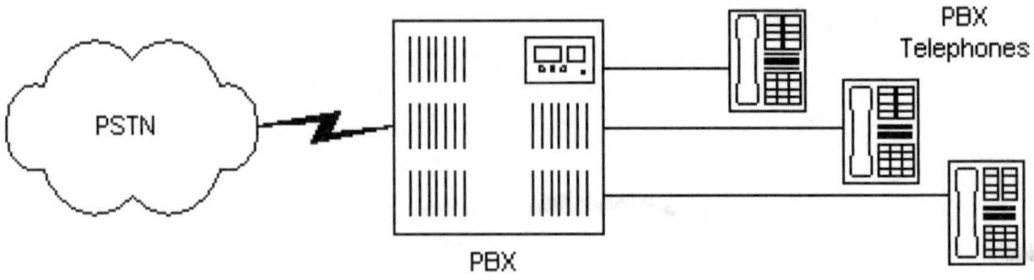

Figure 1-1: A PBX allows users to share a limited number of connections to the PSTN and to call one another without using the PSTN.

The advanced services are often delivered as optional and/or completely separate components. Some of the more common advanced services include:

- Auto-Attendant—An auto-attendant consists of the prompts and menus that a caller hears when a PBX answers an incoming call. The menus usually allow the caller to select a specific person or department. The auto-attendant then transfers the phone call based on the caller's selection. Most auto-attendants sound something like this...

  "Hello and thank you for calling XYZ Corporation. If you know the extension of the person you'd like to speak with enter it now. For sales, press 1. For technical support, press 2. To speak with an operator, press 0."

- Voice Mail—The ability for a caller to leave a recorded voice message when the called party is not available is an optional feature on most PBXs.

- Automatic Call Distribution (ACD)—An ACD system allows calls to be distributed amongst a defined group of users. If all of the users in a group are unavailable, incoming calls will be queued. ACD systems are used by most large call centers. Most of us are familiar with the following message provided by an ACD system...

  "Thank you for calling XYZ Corporation. All of our agents are currently assisting other customers. Your call will be handled in the order that it was received."

- Least Cost Routing (LCR)—LCR service can be used to route outbound calls to different local or long distance service providers based on a variety of factors, such as the time of day or the destination of the call. LCR is designed to take advantage of discounted calling rates offered by service providers.

Today, you will find a PBX installed in almost every office with more than a few employees. Since voice communication is critical to the success of most businesses, the office PBX is an extremely important business tool.

1.3 The PBX Architecture

The traditional PBX has been designed around the characteristics of the PSTN. The PSTN is a circuit switched network that provides a dedicated 64 Kpbs circuit for every point-to-point connection. For example, when you place a phone call from New York City to San Francisco over the PSTN, you reserve a 64 Kbps path between the two locations.

The PBX is simply an extension of this concept. When making a call through a PBX, you are given a dedicated circuit from your telephone through the PBX to the party that you are calling.

The traditional PBX is a specialized computer optimized to provide reliable voice communications. The traditional PBX has the following basic components:

- Cabinet and Backplane—The components of a PBX are contained within a cabinet that is usually designed to be mounted on a wall or on a rack. Inside the cabinet, you will find components that exist in any computer including, a power supply, cooling system, expansion slots, etc.

  A PBX is based on a passive backplane design. All of the main processing and input/output devices are contained on expansion cards that are plugged into slots on the backplane.

- Control System—The operation of a PBX is controlled by a real-time operating system (OS). This OS is often a dialect of UNIX or vendor-proprietary OS. The Central Processing Unit (CPU) of a PBX is contained on an expansion card. Many PBXs have redundant CPUs to ensure greater reliability.

- Switching Matrix—The switching matrix is the component of the PBX that is responsible for creating a voice circuit between an input device and the requested output device. In a digital PBX, time division multiplexing (TDM) switching techniques are used. As with many of the other components of a PBX, the switching matrix is contained on an expansion card.
- Trunk Interface or Trunk Cards- The trunk interface is the component of the system that provides the connection to an external communications network. The trunk interface units are responsible for converting the format and rate of a transmission between that used by the PSTN and that used internally by the PBX.
- Station Interface or Line Cards—The station interface is the component of the system that provides the connection to the terminals.
- Terminals or Station Sets—The telephones connected to the station interface are referred to as terminals or station sets. The telephones of a PBX are often compared to the terminals of a mainframe computer. As with a mainframe computer, the terminals of a PBX are single-purpose, remote input/output devices. PBX terminals vary from simple, single-line analog telephones to proprietary, feature-rich, multi-line phones with built-in display screens. Many PBXs also provide multi-function consoles used by operators or receptionists.
- External Expansion Bus—The expansion bus is the interface of the PBX that is used to connect multiple PBXs together to build a larger system. Often, the expansion bus is used to upgrade the capacity of a system once all of the expansion slots in a single cabinet have been exhausted. Unfortunately, many traditional PBXs do not provide an expansion bus, which is one of the reasons that upgrading a tradition PBX is often referred to as a "forklift upgrade" (i.e. replacing the entire system—cabinet and all).
- Ancillary Application Servers—Ancillary application servers are platforms that run applications such as a voice mail or ACD system. These systems interface to the PBX through a serial link, line cards, or proprietary interface cards.

It is important to note that there is no standard reference architecture for PBXs. Every PBX manufacturer has designed their system to their own proprietary specifications. Accordingly, a telephone designed by one PBX manufacturer will not work with another manufacturer's system—nor will the voice mail unit, the ACD unit, or any other component. In fact, the only standard interface on a traditional PBX is the interface to the PSTN. As a result, the selection of a PBX has traditionally been viewed as the selection of a long-term business partner.

Component	Description
Cabinet and Backplane	Proprietary, mainframe-like cabinet with a passive backplane design.
Control System	An embedded real-time operating system designed specifically for the PBX, often redundant to increase reliability.
Switching Matrix	A time-division or space-division switching matrix that creates a dedicated circuit between an input device and the requested output device.
Trunk Interface	An analog or digital interface to the PSTN.
Station Interface	An analog or proprietary digital interface to PBX terminals.
Terminals	Analog phones, proprietary digital phones, or operator consoles connected to the PBX station interface.
Expansion Bus	A proprietary bus that may require a specific PBX interface card.
Ancillary Application Servers	Single-purpose server (often proprietary) that may require a specific PBX interface card.

Table 1-1: Components of a Traditional PBX

The PC vs. the PBX

One of the primary differences between the Personal Computer (PC) and the PBX is that the PC is designed around a standard reference architecture. It's this standard architecture that allows you to add expansion cards and peripherals from a variety of manufacturers to your desktop PC. This standard architecture has been a major factor in the creation of a dynamic, innovative, and competitive PC marketplace.

The traditional PBX, on the other hand, has not benefited from any standard reference architecture. Each PBX manufacturer has designed a completely closed and proprietary system. Components cannot be shared between systems from different manufacturers and there are no third-party vendors. The result has been a relatively static marketplace with limited innovation and competition—especially when compared to the PC marketplace.

1.4 Media Processing and Call Control

PBX functionality can be split into two main categories: media processing and call control. Media processing specifies how various media formats—voice, fax, etc.—are transmitted by the system. Call control, or signaling, is concerned with activities such as placing, answering, transferring, and conferencing calls.

In a traditional PBX, the media processing is usually limited to voice. Internally, a PBX may transmit voice in a proprietary manner. However the voice signal must be presented in a standard format at the interface to the PSTN.

For a digital PSTN interface, the standard voice coding formats include pulse code modulation (PCM) and adaptive differential pulse code modulation (ADPCM). However, the specific digital coding scheme is usually country or region-specific. The trunk interface unit of the PBX is responsible for converting the transmission format between that required by the PSTN and the internal format of the PBX.

In the same way, the traditional PBX may use a proprietary protocol for internal call control or signaling. However, at the interface to the PSTN, the PBX must conform to certain signaling standards. These standards include POTS (loop start signaling), T-1, E-1, DPNSS, QSIG, and others.

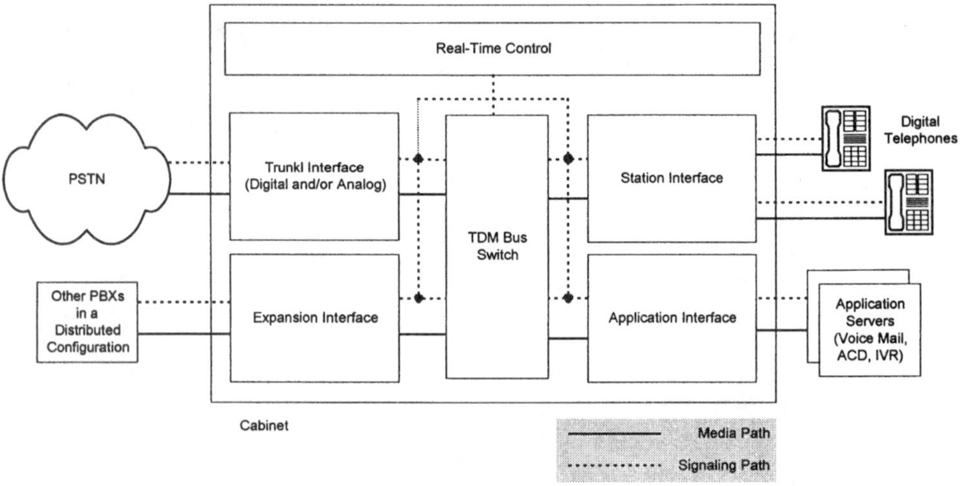

Figure 1-2: Traditional digital PBX architecture based on TDM switching.

PSTN Phone Lines: It's a Jungle Out There

There are so many different types of PSTN phone lines (and acronyms!) that understanding the differences is often a challenge. However, when shopping for a new PBX, understanding the PSTN interface is critical. In a PBX, the trunk interface is specifically designed to integrate with a certain type of PSTN phone line. In other words, you can't just plug any line from the phone company into your PBX and expect it to work. You need to make sure that your phone service is compatible with the trunk interface of your PBX.

The short introduction to phone lines below will help you navigate through some of the terminology you're likely to find when shopping for a new PBX:

- POTS (Plain Old Telephone Service)—POTS lines are ordinary analog phone lines. Chances are that this is the type of service you have in your house.
- Centrex—Centrex refers to a suite of enhanced telephone services, which are typically delivered over POTS lines. Centrex services include; 3 way calling, call transfer, call forwarding, call waiting, voice mail, and more.
- ISDN BRI (Integrated Services Digital Network Basic Rate Interface)—ISDN BRI lines provide two phone lines over a single physical circuit. ISDN BRI service is also referred to as 2B+D, since there are two "bearer" (voice or data) channels and one "data" (signaling) channel. The "data" channel is used to digitally transmit signaling that controls the "bearer" channels.
- T-1 and E-1—T-1 and E-1 are high density digital lines where a single physical circuit uses time division multiplexing (TDM) to handle 24 or 30 channels respectively. T-1 lines are most common in North America and E-1 lines are most common in Europe and Asia.
- ISDN PRI (Integrated Services Digital Network Primary Rate Interface)—ISDN PRI is a service that is run over a T-1 or E-1 line. On a T-1 line, ISDN PRI provides 23B+D service and on an E-1 line, 30B+D service.
- Direct Inward Dialing (DID)—A DID line passes information about the number that was dialed with the call. This information is used by a PBX to route incoming calls to a specific person or department inside of a company. With this service, an external caller can reach an internal extension without going through an attendant. In Europe, this service is also known as DDI.
- Digital Subscriber Line (DSL)—Today, DSL service is primarily used for data transmission and Internet connectivity. However, a number of startups have recently introduced products based on voice-over-DSL (VoDSL) technology.

1.5 Limitations of the Traditional PBX

Today, many organizations use a traditional PBX to provide internal and external voice communication services. The traditional PBX architecture has a number of significant limitations, which have motivated the development of next generation solutions. These limitations include:

- The traditional PBX is a single-purpose communication device optimized to provide reliable voice communication. Just like the Public Switched Telephone Network (PSTN), the design of the traditional PBX is based on the assumption that circuit-switched voice communication is the only type of communication that matters. Yet, with the growth of data networking, and especially the Internet, this assumption is no longer true for many organizations.

- The traditional PBX is designed to interface exclusively to the PSTN for external communications. The digital backbone of the PSTN uses an inflexible audio sample rate of 8KHz (8000 times per second). This is adequate for voice conversations, but it is not appropriate for high-quality audio or other types of multimedia that are becoming increasing common. For example, CD quality audio requires a sample rate of 44KHz.
- The traditional PBX is a closed, proprietary system that cannot be easily integrated with an organization's workflow requirements. Although there have been a number of initiatives to provide a standard PBX programming interface via a CTI link that would allow computer applications to command and control the PBX, the complexity and cost of this approach has hindered it from gaining mainstream acceptance.

Until recently, these limitations were not questioned. They were viewed as a fact of life when dealing with PBXs and voice communication solutions. However, as our communication patterns changed, vendors started to look at voice communication solutions from a fresh perspective—a perspective based on the communication requirements of today, unencumbered by the limitations of the traditional PBX.

1.6 The CTI Link

A number of attempts have been made to provide a more open and standard interface to the closed, proprietary traditional PBX. The term Computer Telephone Integration (CTI) refers to a system architecture where programs running on a standard computer command and control a PBX.

In this architecture, the physical integration between the PBX and the computer (often referred to as a CTI server) is usually accomplished via a serial cable link. The PBX sends status information and event notifications over this link to a program running on the CTI server. In turn, the program responds with commands that control the operation of the PBX.

CTI is used predominately in call centers and other phone-intensive work environments for a wide range of applications designed to increase efficiency and improve customer service. One common application of CTI technology is often referred to as the "screen pop". Such an application allows call center agents to view information about incoming calls on their PCs even before their phone rings.

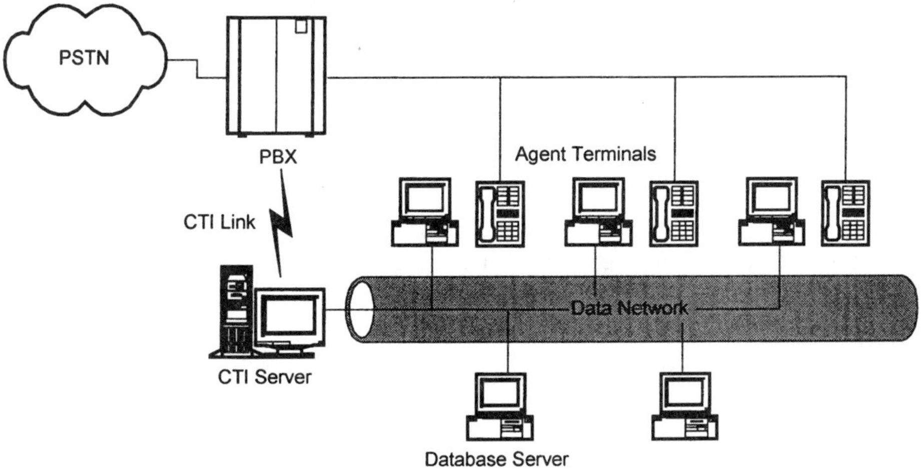

Figure 1-3: CTI Architecture. Applications running on a CTI Server command and control the PBX over a CTI Link.

Here's a simplified explanation of how a typical "screen pop" application works:

1. A customer places a call to a company's call center.
2. Along with the call, the phone company passes some basic information about the call. This information might include Caller ID or Automatic Number Identification (ANI) information.
3. The PBX at the company answers the call and captures the Caller ID or ANI. The PBX sends this information over the CTI link to the CTI server.
4. A program running on the CTI server uses this information to search for data about the caller in the company's database. If the customer had dealings with the company in the past, the company may have information about the caller in their database—indexed by the customer's Caller ID or ANI.
5. If the company has information about the customer in their database, this information can be sent from the CTI Server across the Local Area Network (LAN) to a call center agent's PC. At the same time, the CTI Server can instruct the PBX to transfer the call to the same call center agent.
6. A program running on the agent's PC receives the information and opens a window on the agent's PC screen. When the information appears on the agent's screen it is often referred to as a "screen pop".
7. The call center agent is able to view a history of the caller's relationship with the company and even greet the caller by name when answering the call.

1.6.1 CTI Standards

There have been a number of initiatives to standardize the interface to the PBX over this CTI link. The most notable initiatives have been:

- Computer Supported Telephony Architecture (CSTA)—CSTA is a standard from the European Computer Manufactures Association (ECMA).
- Telephony Server Application Programming Interface (TSAPI)—TSAPI was invented by AT&T and one of the first companies to use the interface was Novell. More recently, TSAPI has become the basis for a call control standard of the Enterprise Computer Telephony Forum (ECTF).
- Telephony Application Programming Interface (TAPI)—The initial specification for TAPI was developed jointly by Microsoft and Intel. Today, Microsoft is driving the evolution of the specification and has incorporated TAPI into its Windows operating systems. It's important to note that the scope of TAPI is much broader than just CTI. TAPI includes all telephony services on the Windows platform—from dialing a modem to real-time voice over IP.

All of these initiatives have enjoyed a certain amount of success at one time or another. However, their success has been limited by the sheer complexity of the implementation. The worlds of telephony and computers have evolved along different paths in terms of technology, distribution channels, and support services. Although the CTI link has been able to bridge this gap in specialized implementations, it has not proved to be an architecture that is suitable for mainstream adoption.

1.7 The Changing Nature of Business Communication

The closed architecture of the PBX and implementation difficulties of integrating this specialized platform with the world of computers and data networking have limited innovation and overall market opportunity. This has become especially apparent in the past few years as data communications and the Internet have exploded into all areas of business and personal communications. Innovations and advancements in communications have been based on the computer and data network platform—not the PBX or even the circuit-switched architecture of the PSTN.

Today, many of us are more likely to send a message by e-mail, than we are to pick up the phone and make a call. Likewise, we are more apt to visit a company's Web site to learn about a new product or service than we are to call the company and request a brochure. While we're visiting the Web site, we might even be able to interact directly with an agent from the company via text-based chat, voice, and/or video over the Internet. None of these innovations are based around the PBX or the architecture of the PSTN.

This shift in the way we communicate is having a significant impact on the role of the PBX within an organization. No longer can the PBX remain a closed, proprietary island in a sea of open, standards-based computing and network resources. The PBX is a critical communications platform that must fully integrate with an organization's overall communications infrastructure and strategy. This requirement is one of the primary forces driving the evolution of the phone system industry today.

1.8 The Enabling Internet

As mentioned above, in the past few years we've seen a fundamental shift away from the services of circuit-switched networks, such as the PSTN, toward the services of packet-switched networks, such as the Internet and corporate intranets. Why are packet-switched networks such a powerful enabling technology that has given rise to so many new and innovative modes of communication? Here's a short list of some of the underlying factors that have motivated this shift to packet-switched network services and applications:

- Packet-switched networks are more efficient. A circuit-switched network dedicates a fixed-width path between end points for the entire duration of the connection. Even if the service running over the network does not require the bandwidth, it is used nonetheless. For example, if both parties on a telephone call are completely silent, the bandwidth of the dedicated circuit is still being used.

  Services running on a packet-switched network use only the bandwidth that they require. The information being transmitted over a packet-switched network is represented digitally and transmitted as small packets. Each packet of a transmission contains its destination address and can be routed independently across the network. At the destination, the packets are reconstructed into their original sequence.

  The services running over such a network can decide exactly what information should be placed in each packet. For example, applications that transmit voice over the network can compress the voice (using algorithms that operate on the digital representation of voice) and eliminate periods of silence.

- Packet-switched networks are more cost-effective. As mentioned above, the PSTN dedicates a circuit between end points of a connection. Accordingly, the use of PSTN resources is easy to measure and users are billed based on the distance and duration of their usage.

  Since a packet-switched network does not create a dedicated end-to-end connection, the concepts of distance and duration are not relevant. The cost basis of a packet-switched network, such as the Internet, is simply bandwidth usage—it costs the same to send an e-mail message across town as it does to send the same message halfway around the world.

- Packet-switched networks are more flexible. The design of the PSTN is optimized for reliable, quality voice communication. A variety of assumptions have been designed into the network itself that limit the ability of the PSTN to be used for other services. As stated previously, the PSTN uses an inflexible sample rate of 8KHz. This is fine for voice conversations, but it is not appropriate for high-quality audio (for example, CD quality audio requires a sample rate of 44KHz).

The Internet, on the other hand, is designed to transmit packets of data without making any assumptions about what the packets might contain. It is up to the intelligent applications running on the computers at the end points of the connection to process the information. This flexibility has made the Internet an attractive platform for software developers who are building a wide range of innovative communication applications.

The popularity of the Internet and data communications in general would not have been possible without the PC. When the PSTN was designed in the early 60's, even the earliest model of a PC was more than a decade away. Accordingly, the PSTN was designed to work with relatively dumb, cost-effective terminals (i.e. telephones) and place the "intelligence" in the network itself.

Now, almost 40 years later, PCs are outselling even television sets in the United States. As the processing power that PCs deliver becomes more commonplace, so too are the communication services that rely on this power and take advantage of the flexibility offered by a packet-switched network.

The Internet vs. The PSTN

To understand the Internet and the role it plays in supporting a wide variety of applications, it's necessary to be familiar with the fundamental principles behind the Internet and how it compares to the PSTN. Although the Internet shares some characteristics of the PSTN, it also has significant differences.

The PSTN is a circuit-switched network that has been optimized for real-time or synchronous voice communication with a guaranteed quality of service (QoS). When a telephone call is initiated, a circuit is established between the calling party and the called party. The PSTN guarantees the QoS by dedicating a full-duplex 64Kbps circuit between the parties of a telephone conversation. Regardless of whether the parties are speaking or silent, they are using a 64Kbps dedicated circuit until the call ends. Since the bandwidth remains constant, the cost of a telephone call on the PSTN is based on distance and time.

On the other hand, the Internet is a packet-switched network that has historically been used for applications where maintaining a real-time QoS is not necessary, such as e-mail and file transfers. Packet-switched networks do not dedicate a path between sender and receiver and therefore cannot guarantee a QoS for real-time applications. (There are, however, initiatives currently underway aimed at providing QoS on packet-switched networks, such as the Internet. See *The Quality of Service Debate* below).

How does a packet-switched network like the Internet work? Conceptually, it's quite simple. All information to be transmitted over the Internet (voice, text, images, etc.) is divided into packets that contain a destination address and a sequence number. Internet routers and servers direct these packets over the Internet until they arrive at their destination. From the perspective of the Internet all packets are treated exactly the same regardless of their content.

Once packets begin to arrive at their destination, the sequence number is used to recompose the packets into their original order. For applications where real-time interaction is not necessary, such as e-mail, the order in which the packets arrive or the delay between packets is not tremendously significant. However, real-time applications, like telephony, can degrade considerably when there are long delays between the arrival of packets. For this reason, packet-switched networks have historically been thought of as inappropriate for real-time applications like telephony.

The primary cost mechanism for Internet usage is not distance or time, but bandwidth usage. Although many Internet Service Providers (ISPs) currently offer consumers unlimited usage of the Internet for a fixed monthly fee, businesses and ISPs themselves are often billed for Internet usage based on bandwidth.

Although today's Internet does not guarantee QoS for real-time applications, that hasn't stopped it from becoming popular for voice communication. Callers are more than willing to tolerate a slight amount of delay or reduced voice quality in exchange for inexpensive long distance voice communication. With the current price structure of the Internet, Internet telephony users can speak with someone anywhere in the world often for the price of a local phone call.

The Quality of Service (QoS) Debate

The PSTN and the devices that are designed around the PSTN architecture, such as the PBX, are optimized to deliver quality voice communication by dedicating bandwidth between the end points of a connection.

On most of the Internet today, bandwidth is not dedicated or reserved for any application. All of the data packets that are sent across the Internet are treated equally. Packets of a real-time voice conversation are given the same priority as less time-sensitive data such as e-mail messages.

The effect is that real-time applications, such as voice communication, do not deliver the same QoS over the Internet as they do over the PSTN. At best, the quality today is variable.

However, no discussion of QoS would be complete without mentioning the following important points:

- The Internet is a "network of networks" with no centralized management. Accordingly, when transmitting information across the Internet it is not always possible to know what segments of the Internet will be used and how well they are managed. Transmitting information over a private packet-switched network, such as a corporate intranet, is entirely different. An intranet can be managed centrally and provisions can be made to assure that there is sufficient bandwidth for real-time applications.
- A number of bandwidth reservation protocols and priority queuing techniques are in the process of being implemented throughout both corporate intranets and the Internet. This should result in networks that can provide a quality of service that is appropriate for the application.
- The amount of bandwidth on the Internet continues to increase. Everyday, backbone carriers and Internet Service Providers (ISPs) add bandwidth to increase the capacity of their networks. With technologies that deliver more bits per second, such as optical networking, it might just be possible that the whole QoS discussion will be mute in a few short years.

1.9 Summary

The PBX is a critical communication platform for the vast majority of businesses today. The changing nature of business communication and the limitations of the traditional PBX have given rise to an entirely new class of communication solutions. These solutions are based on innovations in the computer and networking industries and build on the open, standards-based foundations of these industries. We refer to these solutions as next generation phone systems and they are the focus of the chapters that follow.

Chapter 2

Overview of Next Generation Solutions

2.1 Introduction

In the past few years, a variety of new voice communication solutions have been introduced to the market. In contrast to the previous generation of closed, proprietary PBX systems, these new solutions take advantage of open, standard computer and networking technologies. To identify these solutions and differentiate them from their predecessors, they are commonly referred to as "next generation phone systems". You may also hear a variety of other terms used to refer to these solutions, including "PC-PBXs", "IP-PBXs", and "Communication Servers". As we will see, each of these terms refers to a more specific type of next generation phone system. The common thread of all next generation solutions is that they are designed to take advantage of the convergence of data, voice, and video.

There is no single architecture that is common to all next generation phone systems. Each next generation solution leverages computer and networking technologies in different ways and to different degrees. Today, popular next generation approaches range from circuit-switched PC-based phone systems to network-based solutions that provide packet-switched voice to the desktop—and many variations in between. In the chapters that follow, we will explore each approach in detail.

2.2 State of the Market

Next generation phone systems are positioned to compete directly with the traditional PBX as the voice communication platform within an organization. As the market for next generation solutions evolves, there are a number of important trends that define the state of the market. The most important trends are:

- During the past few years, not only new products but also entirely new architectures have been introduced to the market on a regular basis. Although this can make for an interesting and exciting marketplace, it can also be disorienting for potential customers, as they weigh the pros and cons of different architectures and feature sets.
- Consolidation of a variety of communications services—voice, fax, data access, etc.—into a single platform is a common design goal of many next generation solutions.
- Support for packet-switched media is playing an increasingly important role in next generation solutions. Convergence of voice and data into a single network infrastructure is a primary design goal of many next generation solutions.
- Most next generation solutions incorporate a variety of advanced phone system features as part of the standard offering. These features often include auto-attendant, voice mail, automated call distribution (ACD), follow-me call forwarding, desktop call control, unified messaging, least cost routing, direct programming interfaces, and more.
- Next generation solutions are based on a variety of established and emerging standards that govern basic media processing and call control functions. These standards include ECTF S.100, ECTF S.300, ECTF H.100, ECTF H.110, ECTF A.100, ECTF A.130, ECTF C.100, ITU H.323, ITU T.120, IETF MGCP, IETF SIP, and Microsoft TAPI.
- Next generation solutions began arriving on the market in the mid-1990s. Accordingly, most solutions on the market today are in their first or second product release cycle. The most mature solutions on the market are only in their third or fourth release cycle.
- Many traditional PBX manufacturers have also announced next generation strategies or products that take advantage of the convergence of data and voice communications.

The Importance of Standards

Why are standards important? Standards are important because they provide customers with the flexibility to integrate components from multiple vendors into a single solution. Standards allow customers to design multi-vendor solutions that meet their specific needs and fit their budgets. At the same time, standards encourage vendor specialization and promote a more competitive, innovative marketplace.

Of course, there is often a price to pay for achieving standards-based interoperability. Until standards mature and true plug-and-play interoperability is achieved, the market often experiences growing pains as components don't always integrate as they should. When the integration pains outweigh the benefits of interoperability, standards are often forgotten.

One of the best examples of both the advantages and disadvantages of standards-based interoperability is the personal computer industry—just compare the standards-based IBM-compatible PC with the Apple Macintosh. The market share advantage of the IBM-compatible PC is due in large part to an open reference architecture that fostered competition, innovation, and customer choice. Even though the IBM-compatible PC industry was plagued by interoperability issues for many years, the advantage of a open, public reference architecture outweighed the disadvantages and helped the IBM-compatible PC achieve greater market share than the proprietary Apple Macintosh platform.

2.3 Features and Benefits

Why are next generation phone systems receiving so much attention? Here are some of the features and benefits that are making next generation phone systems increasingly popular:

- Ease-of-use—When compared to traditional PBXs, next generation solutions set a new standard for usability. Next generation phone systems can be configured and administered with a familiar graphical user interface, such as Microsoft Windows. No more proprietary, character based interfaces. Now, just point, click, drag, and drop.
- Cost—With a modular design and reliance on generic components, next generation solutions compete very favorably with proprietary platforms on price. When the complete complement of technologies and enhanced features is considered, next generation solutions are often a fraction of the cost of their traditional counterparts.
- Expandability—The computer and communications industries are moving at an exciting pace. New services and applications are constantly being introduced. Closed, proprietary platforms won't support these developments. Open, next generation platforms are quick to embrace these technologies. For example, many next generation solutions offer Web-based interfaces for administration and account management—something rarely found on a traditional PBX platform.

- Consolidation of Infrastructure and Administration—Many next generation solutions use an organization's existing data networking infrastructure to transmit voice and/or video. One of the most important benefits of next generation solutions is the ability to consolidate all forms of communication on a single network infrastructure. The result can be significant cost savings and elimination of administrative duplication.
- Support for Desktop CTI—Next generation phone systems directly integrate with an organization's LAN. Most include their own desktop CTI software and/or support for a client/server CTI interface, such as TAPI.
- Customization—Many next generation phone systems include an application programming interface (API). Accordingly, if an organization requires a highly customized phone system solution, they can develop it themselves. In fact, some next generation solutions are primarily sold as a platform upon which customized applications can be developed.
- Fault Tolerance—If a traditional PBX fails, you wait for a specialist to come on-site to fix it. Many next generation phone systems are built on a standard computing platform, such as Windows NT. If the system does fail most organizations have administrators who can reboot the PC, run standard diagnostics, and troubleshoot the system. Other next generation solutions are designed in a distributed manner and eliminate a central point of failure. If any of the distributed components fail, they can be easily replaced without causing a service outage to the entire system.
- Features, Features, Features—Next generation phone systems provide much more than basic telephony services. Standard features often include unified messaging, ACD, voice mail, follow-me roaming, auto-attendants, caller identification and screening, remote configuration via the Web, Internet services, and more.

2.4 Shortcomings

Next generation solutions offer many features and benefits that can't be matched by their traditional PBX counterparts. Traditional PBX manufacturers, however, are quick to highlight deficiencies of next generation solutions, especially in such critical areas as reliability and scalability. Although the accuracy of these claims is often a source of heated debate, there are a number of areas where next generation solutions sometimes fall short in comparison to traditional PBXs. These areas include:

- Proven Reliability—Voice communication is a mission-critical application. The traditional PBX is a mature solution that has proven its reliability time and time again. Next generation solutions are relatively new to the market and many do not yet have a proven track record. This is not to imply that next generation solutions are inherently less reliable then traditional PBXs. In fact, many next generation solutions are specifically designed to meet the mission-critical needs of call centers and other phone-intensive workgroups. However, some next generation solutions have simply not yet *proven* that they are as robust as traditional solutions.

With each passing day, any criticism of the reliability of next generation solutions becomes less valid as an increasing number of next generation solutions are being deployed in mission-critical applications.

Next Generation Personal Computers

To one degree or another, PCs are used as a component in almost all next generation voice communication solutions. Early on, critics of next generation solutions were quick to point out that the PBX was based on a more reliable architecture than that of the standard PC—and therefore provided a more reliable voice communications platform.

One of the subtle architectural differences between the standard PC and the PBX is that the PBX uses a *passive backplane* design, while the standard PC does not. In a standard PC, the motherboard contains the Central Processing Unit (CPU), system memory, and other components. In a PBX, the concept of a motherboard does not exist. A PBX contains a backplane with expansion slots and all of the system components are contained on expansion cards. With such an architecture, it's possible to increase reliability with redundant expansion cards. Additionally, expansion cards in most PBXs are *hot-swappable*. That is, they can be removed and replaced while the PBX is operating.

Some of the early implementations of PC-based telephony solutions were plagued with accusations of unreliability based on the fact that the standard PC did not provide a passive backplane that supported hot-swappable components. However, as more and more mission critical applications, such as telephony, began moving to the PC platform, PC manufacturers responded. They designed systems with passive backplanes and expansion buses based on the Compact Peripheral Component Interface (CompactPCI), which supports hot-swappable components. These PCs, referred to as fault-resilient or industrial PCs, have silenced most critics of PC-based telephony systems.

- Scalability—Most next generation solutions are designed for small to medium sized businesses or workgroups of up to a few hundred users. Within this segment of the market, increasing system capacity can usually be accomplished quite easily by adding an expansion card to a PC or by plugging a turnkey appliance into a LAN. However, today, few next generation solutions are designed to address the enterprise market.

For most next generation phone system vendors, scalability is a top priority. With each new product release cycle, next generation solutions are moving closer to the scalability of the traditional PBX platform. Even today, you can find PC-based solutions that scales to 500+ users, distributed solutions that support thousands of users, and IP-based solutions that scale up to 10,000+ users.

- Worldwide Regulatory Approval—In most countries any product or system that will connect to the PSTN must be formally approved by the appropriate regulatory authority. As this can be an expensive and lengthy process, it forms an initial barrier of entry for some next generation solutions in some countries. The leading traditional PBXs have regulatory approval in most countries around the world.
- Trunk Interface Support—Today, many next generation solutions do not provide support for a wide variety of trunk interfaces. With most next generation solutions in only their first or second release cycle, the design targeted a specific market in a specific country or region, with the target market usually implying a specific type of trunk interface. For example, many of the North American vendors initially supported analog trunk line interfaces, while many European vendors initially supported ISDN trunk line interfaces. As vendors expand their markets, they will also provide support for additional trunk line interfaces.
- Feature Phone Support—Although next generation solutions usually offer a broader range of enhanced features than traditional PBXs, there is one area where traditional PBXs have an advantage—the feature phone. Many next generation solutions are designed to only work with standard analog telephones. Traditional PBXs, on the other hand, usually include sophisticated, multi-line telephones with built-in display screens, special feature keys, support for call appearances (so that co-workers can see the status of your extension and even answer your calls if you're already on the phone), and more. Next generation solutions typically provide these features through a software application that runs on a user's PC. While this approach may be sufficient for many users, it does not take into account non-PC literate users or users who many not have or need a desktop PC.

  Although many next generation solutions can't match the sophistication of traditional PBX feature phones, there are an increasing number that can. This is especially true of solutions that deliver packet-switched voice directly to the desktop. In fact, there is a burgeoning industry of third-party companies, such as Circa Communications and Pingtel, that focus exclusively on developing standards-based feature phones for next generation phone systems. The fact that next generation systems are based on open standards will lead to increased variety and sophistication of desktop telephones in the coming years.

- Novelty—The fact that many next generation solutions are new to the market and manufactured by relatively small organizations is a concern for many potential buyers. It's often easier to justify the purchase of a traditional PBX system manufactured by a company with a well-known reputation and worldwide brand recognition than it is to purchase a next generation solution created by a twenty-five person startup in their first product release cycle. It used to be said that no one was ever fired for buying computer hardware from IBM. When it comes to making a strategic decision about a voice communication solution, the same can be said about traditional PBX manufacturers.

  As a result, it's no surprise that there has been a fair amount of merger and acquisition activity in the next generation phone system industry. Established organizations have been quick to target hot startups—and the startups have been open to the advances. Companies such as Cisco, 3Com, Lucent, and Ericsson have all acquired next generation phone system vendors. These organizations have the resources to invest in long-term product development and the reputation to make potential buyers feel comfortable.

2.5 Summary

The next generation phone system market is extremely dynamic. Over the past few years, this market has tracked the rapid pace of development in the computing and networking industries. The result has been an explosion of next generation solutions and architectures. Two of the most noticeable trends in the market are the migration from circuit-switched technologies to packet-switched technologies and the consolidation of a variety of communication services into a single platform.

Although next generation solutions are packed with an unmatched array of features, they sometimes do exhibit shortcomings when compared to traditional PBXs. Most of these shortcomings can be attributed to the newness of next generation solutions, rather than any inherent deficiency.

In order to stay competitive with next generation solutions, many traditional PBX manufacturers have announced next generation migration strategies. At the same time, many established voice and data communications companies have acquired next generation startups.

Chapter 3

Classification of Next Generation Solutions

3.1 Introduction

As mentioned in Chapter 2, the term "Next Generation Phone Systems" refers to a class of communication solutions that take advantage of the convergence of data, voice, and video. The variety and diversity of these solutions make it challenging to establish a consistent system of classification for the purpose of understanding and comparing solutions. Yet, in order to provide potential buyers with a framework for understanding the market, a classification scheme is required. This chapter introduces a classification scheme and identifies specific categories of solutions that will be used for organizing the remainder of the book.

The goal of the classification scheme presented below is to help you understand which characteristics of next generation solutions are most important when selecting a new phone system. Within this context, you will be able to make an "apples to apples" comparison of solutions and accurately select the appropriate phone system for your organization.

3.2 Primary Characteristics

The following characteristics of next generation solutions are the most important when classifying the solutions in the market. These characteristics define the strategic orientation of a solution and can be used as a high-level framework for comparing and contrasting the next generation phone systems available today.

3.2.1 Infrastructure Requirements and Support for Convergence

When we consider infrastructure requirements we are primarily interested in understanding whether a solution runs on a dedicated infrastructure or a converged infrastructure. In other words, to what extent will a particular phone system eliminate the need to manage and administer separate voice and data networks?

Maintaining a dedicated voice communication infrastructure is expensive, but until recently most organizations didn't have an alternative. Traditional PBXs run on proprietary platforms that require their own wiring and physical plant. In addition, traditional PBXs are often supported and administered by trained personnel who specialize in a particular system. Today, most organizations have two parallel infrastructures for voice and data networking. The consolidation of these infrastructures can lead to cost savings, streamlined personnel requirements, and more flexible administration and management.

Infrastructure convergence is one of the primary benefits of next generation solutions and it's one of the main reasons that organizations view next generation solutions as a strategic alternative to the traditional PBX. Accordingly, it is one of the primary characteristics that we will look at when we categorize and compare next generation phone systems.

Full infrastructure convergence is made possible by phone systems that support real-time, packet-switched voice transmission (i.e. transmitting voice over the data network infrastructure). As we'll see, next generation solutions take advantage of the data network infrastructure for voice transmission in a variety of different ways. Some solutions transmit packet-switched media directly to the desktop, while others use the data network as an expansion bus—a means of linking multiple units together to form a larger system. In addition, a degree of infrastructure convergence is also achieved with solutions that consolidate voice and data access into a single physical platform. Typically, these solutions still use separate networks for voice and data transmission. However, they do offer consolidation of the hardware and software into a single box, with a single administration interface.

The degree to which a solution requires its own infrastructure is an important characteristic when comparing next generation phone systems. It's equally important to understand if a solution provides a migration path to a converged infrastructure or if it completely replaces an existing voice communication infrastructure. It's critical to align your organization with a manufacturer whose vision for converged voice and data networks is consistent with your own organization's strategy.

Circuit-Switching vs. Packet-Switching

The switching technology of a next generation phone system defines how real-time media—voice, video, etc.—is carried between the components that make up the overall solution. The switching technology is either circuit-switched or packet-switched and it is the main characteristic that determines the infrastructure requirements.

One of the primary differences between a circuit-switched network, such as the PSTN, and a packet-switched network, such as the Internet, is that a circuit-switched network establishes a dedicated path between the communicating devices. Packet-switched networks, on the other hand, do not establish a dedicated or reserved path between the communicating devices. Rather, information is transmitted as a series of packets, each of which can take an independent path to the destination. The connection between the source and destination nodes in a packet-switched network is often referred to as a virtual circuit.

The switching technology of a next generation solution should be considered for a variety of interfaces, including the station interface, the trunk interface, and the external expansion bus interface. Some solutions support both circuit-switching and packet-switching technologies and/or contain a gateway for converting media between circuit-switched and packet-switched networks.

3.2.2 System Integration Architectures and Interfaces

The system integration architecture of a next generation phone system is concerned with how the components of an overall solution are integrated and configured. For example, does the solution require specialized computer telephony hardware and software to be installed directly into a PC? Or, does the solution consist of network appliances and devices that are connected directly to a data network via RJ-45 connectors? Two of the most common integration architectures of next generation phone systems are PC-integrated solutions and network-integrated solutions.

PC-integrated solutions are phone systems where all of the components are integrated at the level of the PC itself. That is, all of the components of the system—interface cards, system control, applications—are integrated into a standard PC platform (see Figure 3-1). The system is setup and enhanced by installing hardware and/or software directly into a standard PC. The telephony hardware conforms to the ISA, PCI, or CompactPCI bus standard and the software is designed to run on a mainstream operating system such as Windows NT or Linux.

Network-integrated solutions consist of network appliances and PC applications that are integrated at the level of the network. Whereas all of the components of a PC-integrated solution reside in a centralized PC, the components of a network-integrated solution may be distributed throughout the network (see Figure 3-2). For example, the control system may be a PC application that resides on a different platform from the switching system or the PSTN interface. In this architecture, the packet-switched data network is used as the system bus to transfer both call control information and real-time media.

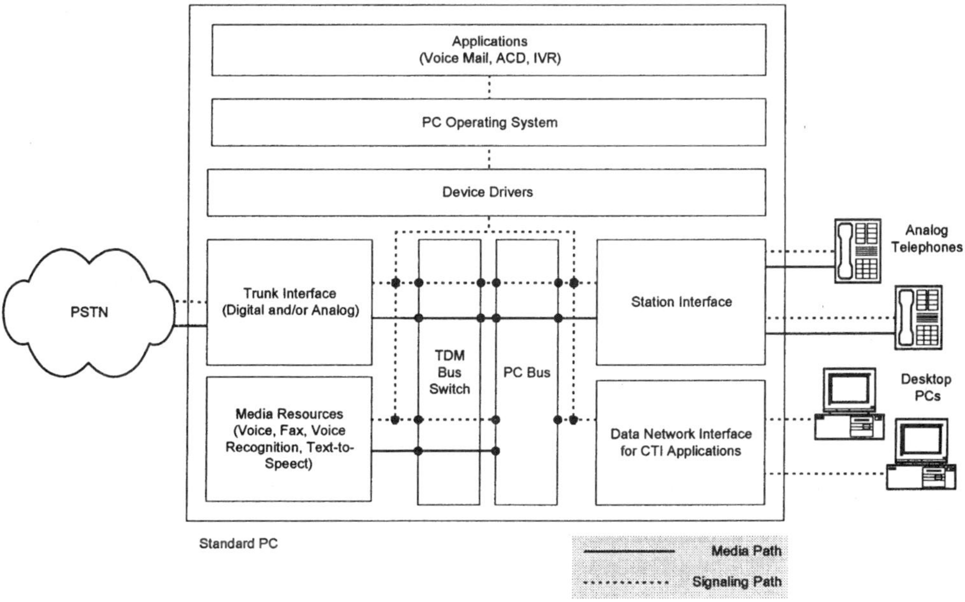

Figure 3-1: Architecture of a typical PC-integrated phone system.

When considering the system architecture, it is also important to consider the interfaces between the system components. One of the main limitations of the traditional PBX is that it is a closed, proprietary platform. With the exception of the interface to the PSTN all of the interfaces of a traditional PBX are proprietary. It's unthinkable to use one manufacturer's phones with another manufacturer's station cards. Even more remote is the possibility of using one vendor's expansion cards in another vendor's chassis. The promise of next generation phone systems is standard integration interfaces that support multi-vendor solutions.

Systems built on open platforms with standard interfaces will reduce your dependency on a single supplier and give you the flexibility to mix and match phone system components and applications from multiple vendors. The industry has yet to fully realize this vision, but—thanks to standards organizations such as the ECTF, IETF, and the ITU—standards do exist and they are now being implemented. Only with widely supported standards for call control, media processing, and hardware component integration will true multi-vendor solutions be possible.

Both the system architecture and the degree to which a next generation phone system supports open integration interfaces are important considerations when comparing solutions. Since many of the core standards for these interfaces are just emerging, it's equally important to understand a manufacturer's commitment to implementing and supporting these interfaces in the future.

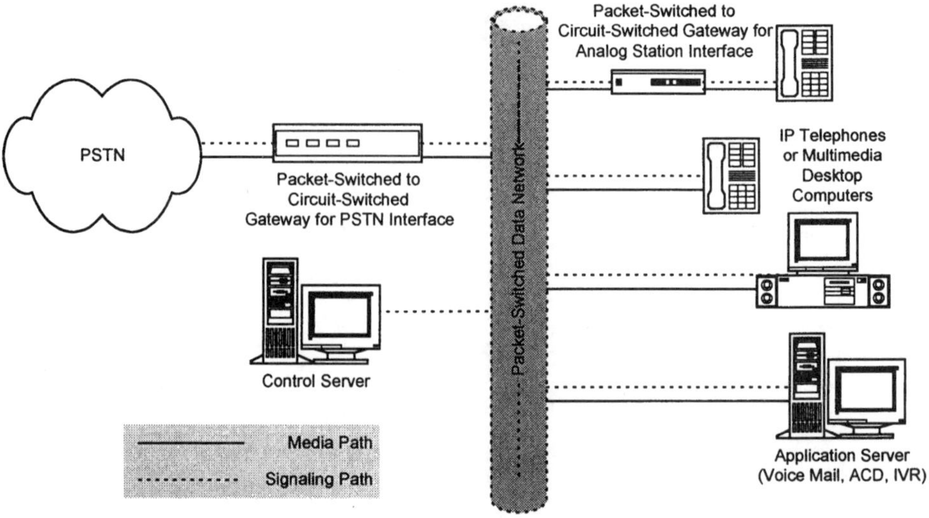

Figure 3-2: Architecture of a typical network-integrated phone system with system components distributed throughout the network.

What is an "Open" Platform?

What are the best criteria to evaluate the openness of a next generation phone system platform? Is a solution based on a standard PC more "open" than a solution based on a proprietary network appliance? The answer to this question depends on what we mean by "open". An open solution is one that is composed of components that have non-proprietary interfaces—a solution that does not lock you into a specific vendor's products. For example, a PC-based solution that requires proprietary telephony expansion cards may be less open than a network-based solution that relies on standard network routers and switches—even though the hardware of the network routers and switches may be proprietary. When looking for an open solution, look for non-proprietary interfaces.

3.2.3 Vendor Focus

Although there is no guarantee that the solution you buy today will continue to meet the future needs of your business, one way to protect your investment is to align yourself with a manufacturer and a solution that is oriented toward the specific needs of your type of organization. Most vendors target small companies, branch offices, workgroups, or enterprises. Vendors design their solutions with the appropriate features, applications, and scalability requirements to support the market they are targeting.

Next generation solutions often provide support for a wide range of enhanced applications, including auto-attendant, ACD, voice mail, follow-me call forwarding, wireless terminals, integrated data access, LAN switching, and more. The specific feature set and the way the features are delivered—as standard, optional, or custom-built components—depend on the market that a vendor is focused on.

It is important to select a vendor that is committed to providing solutions for your segment of the market—today and in the future. A vendor's target market should be consistent with their current product offering, their future release plans, and their strategic vision.

3.3 Market Overview

When we apply our classification system to the solutions in the marketplace, we see that most of the next generation solutions begin to "cluster" in three distinct groups. These groups can be characterized as: PC-PBXs, Communication Servers, and IP-PBXs. Figure 3-3 provides a relative comparison of these solution categories using the characteristics discussed above. Each of these categories is introduced below and covered in extensive detail in the chapters that follow.

3.3.1 PC-PBXs

PC-PBXs were the first of the next generation solutions to appear on the market. The very first PC-PBXs were MS-DOS based solutions that were introduced to the market in the early 1990s. Although these solutions met with moderate success, interest in PC-PBXs increased dramatically in 1996 with the introduction of Windows NT and UNIX-based systems. More recently, a number of vendors have also announced plans to support Linux, the popular open source version of UNIX.

The main distinguishing features of PC-PBXs are:

1. All system components of a PC-PBX are integrated in a standard PC. In addition, there is an active standards organization, the Enterprise Computer Telephony Forum (ECTF), working to develop open interfaces between the components of a PC-PBX. The ECTF is developing a standard reference architecture for a PC-PBX that can support multi-vendor solutions. There is no standards initiative for traditional PBXs that matches the work of the ECTF in terms of scope or vision.
2. Like the traditional PBX these solutions rely primarily on circuit-switched technology for voice transport. However, most PC-PBXs provide a migration path to converged networks by offering gateways that allow voice to be transmitted over packet-switched networks.

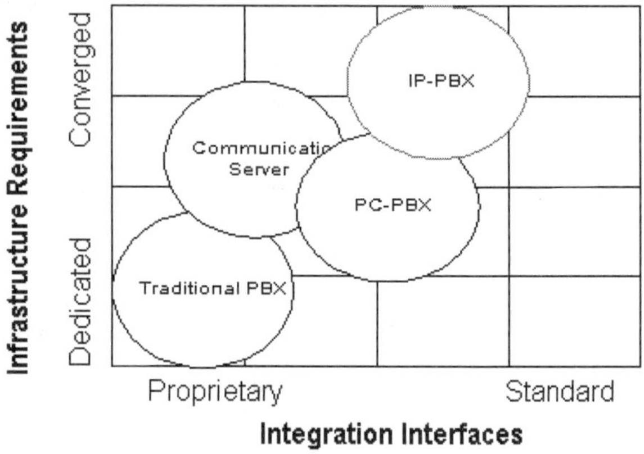

Figure 3-3: Comparison of next generation phone systems. (Note: this diagram provides a snap-shot of the market today. It is a generalization designed to give the reader a basic framework for comparing different solution categories. Markets and products change rapidly and individual products should be evaluated independently using the criteria established in Section 3.2.).

3. PC-PBXs exploit the programmability of the standard PC platform to provide an unmatched range of advanced features and applications, such as auto-attendant, ACD, voice mail, follow-me roaming, CTI, fax integration, unified messaging, and more. Unlike traditional PBXs, PC-PBXs are programmable application platforms.

4. PC-PBXs are usually targeted to small to medium size enterprises or communication-focused workgroups, such as a group of call center agents. PC-PBXs are often deployed "behind" a traditional PBX to provide advanced communication features to a specific workgroup.

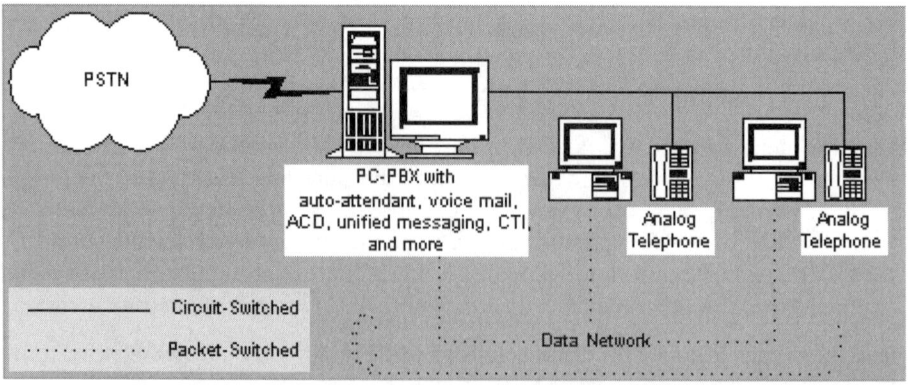

Figure 3-4: The PC-PBX as the voice communication hub of a small to medium sized enterprise.

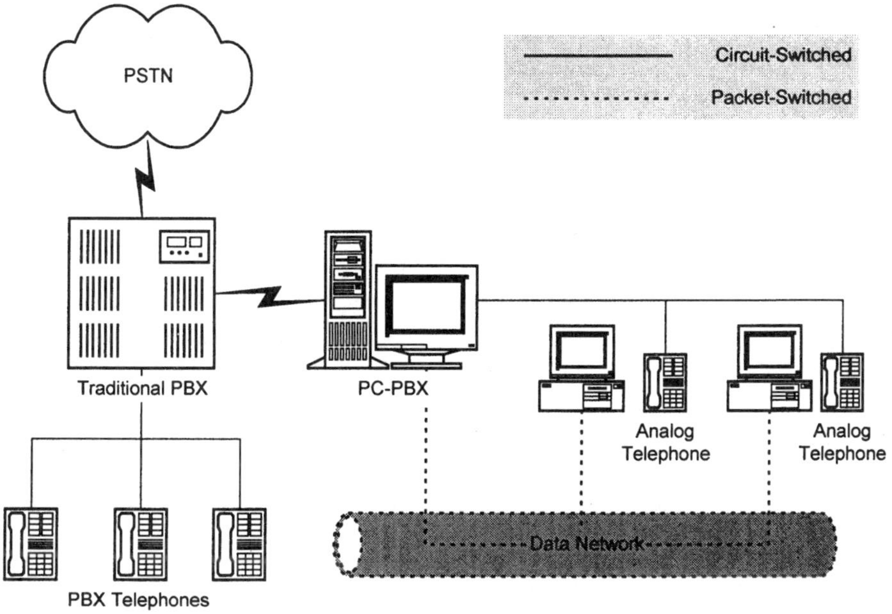

Figure 3-5: Using a PC-PBX with a traditional PBX to provide enhanced services to a specific workgroup of telephone-intensive users.

The Multi-Media Call Center and the PC-BPX

At one time, call centers communicated with customers exclusively over circuit-switched PSTN phone calls. The explosion of the Internet quickly changed that. The new communication alternatives made possible by the Internet—e-mail, text chat, voice/video over IP, etc.—have forced call centers to be more flexible. If a call center is not prepared to handle Internet-initiated inquires, their customers may find it more convenient to work with a competing organization that is prepared.

Accordingly, call centers have had to evolve from handling a single type of media—circuit-switched voice—to handling multiple media types. Hence, the birth of the multi-media call centers. Multi-media call centers usually employ sophisticated ACD solutions that are designed to simultaneously queue and distribute requests in a variety of media types.

Because of their support for a wide range of applications and features, PC-PBXs are an ideal platform for handling the requirements of multi-media call centers. In fact, a number of PC-PBXs, such as Interactive Intelligence's Enterprise Interaction Center, are designed specifically to address this market.

3.3.2 Communication Servers

Communication Servers were first introduced to the market in the late 1990s. The primary design goal of these solutions is to consolidate access to a variety of media—typically voice and data—from a single device.

The main distinguishing features of Communication Servers are:

1. All of the components of a Communication Server are usually contained within a turnkey network appliance. Scalability is often accomplished with a proprietary expansion bus interface or by adding hardware modules to the appliance. Enhanced applications, such as voice mail, are usually contained within the appliance, but they may also run on an external PC server.
2. Communication Servers are primarily based on circuit-switched technology. However, some of these solutions offer a gateway to integrate with packet-switched voice networks.
3. Communication Servers usually provide access to both voice and data networks from the same platform. In addition to providing a complete phone system (usually with auto-attendant, voice mail, ACD, unified messaging, CTI, and more), Communication Servers also include a data network router and hub. Often Communication Servers exclusively support digital trunk interface lines (such as ISDN) so that a single interface can be used for voice and high-speed data access.
4. Communication Servers are usually targeted at small to medium size enterprises or branch offices that are looking for enhanced features in a turnkey system. By consolidating a variety of communication services into a single platform, the installation and administration is more straightforward than a solution composed of multiple independent systems.

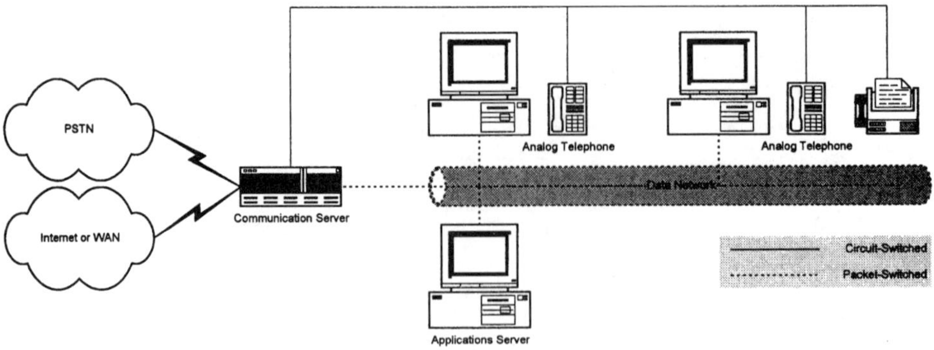

Figure 3-6: Communication Servers are turnkey appliances the often provide access to data and voice communication services.

3.3.3 IP-PBXs

IP-PBXs also began appearing on the market in the late 1990s. Most of these systems were designed from the start to replace the circuit-switched voice communication infrastructure. In general, all of these solutions rely on packet-switched voice technology for one or more of the main system interfaces—the trunk interface, the station interface, or the expansion bus interface.

The main distinguishing features of IP-PBXs are:

1. The components of an IP-PBX solution are usually distributed throughout a packet-switched network. Although most of these solutions include turnkey appliances that are integrated into an organization's data network, some of these solutions simply piggy-back on an organization's existing data network infrastructure. Often these solutions include special telephones that handle packet-switched voice delivered directly to the desktop. The interface between the components of an IP-PBX is the packet-switched data network. A number of standards, such as ITU H.323 and SIP, are designed to provide the open call control and media processing interfaces between these components.

2. IP-PBXs treat real-time voice communication as an application that runs over an existing data network infrastructure. Since these systems carry voice calls over the data network, they eliminate the need for an organization to operate separate voice and data networks. There is typically no centralized circuit-switched platform in an IP-PBX. Circuit-switching is only employed within gateways to provide an interface to the "legacy" PSTN to support analog station sets.

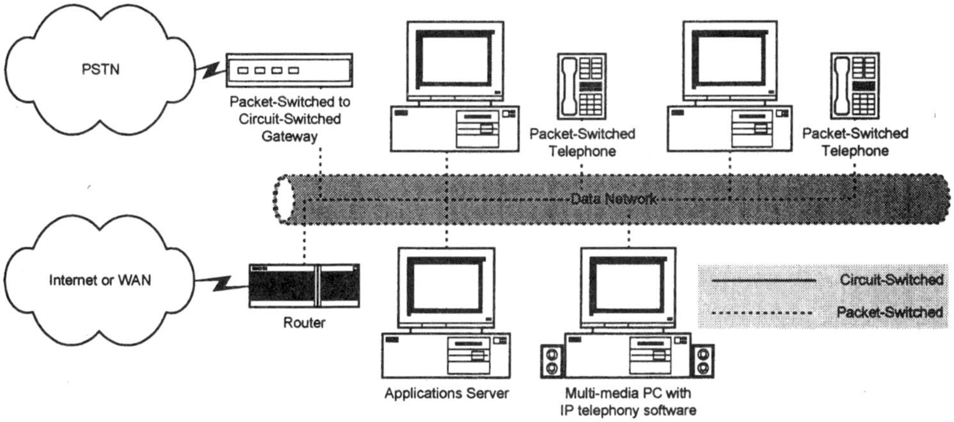

Figure 3-7: With an IP-PBX, voice communication uses the same switching infrastructure as the data network.

3. Although IP-PBXs do provide many enhanced phone system applications, the range of enhanced features is usually not as complete as that offered by PC-PBXs or Communication Servers.
4. IP-PBXs are often designed as enterprise solutions (Cisco, for example, claims their IP-PBX will support over 10,000 users). However, IP-PBX vendors realize that many organizations require a migration path to a full-blown converged network. Therefore, IP-PBX solutions are often positioned as branch office solutions—a way of consolidating the infrastructure in a branch office before ultimately linking all of the office together in an enterprise system. Provided that the data network is managed correctly, these solutions are extremely scalable.

The classes of phone systems introduced in this section are discussed in more detail in the chapters that follow. The table below provides a summary of the components of next generation phone systems.

3.3.4 Virtual Phone Systems

In addition to physical, on-site phone systems, many small to medium size organizations opt for a virtual phone system. What is a virtual phone system? Virtual phone systems replace customer premise equipment with intelligent services that reside inside the public telephone network. The simplest implementation of the virtual phone system concept is Centrex service offered by local phone companies. Organizations can subscribe to Centrex service and receive advanced features—call forwarding, call transfer, call waiting, conferencing, speed dialing, least cost routing, etc.—all on their single line phones without the need for a centralized, on-premise phone system. In effect, Centrex customers are remote users of the CO's feature-rich telephone switch.

Today, virtual phone systems are available that do much more than the local phone company's Centrex service. The virtual phone systems introduced to the market in the past few years provide all of the features of Centrex, plus unified messaging services, Web-based configuration and administration, ACD, voice recognition, and more. As the Application Service Provider (ASP) model becomes more commonplace for other business applications, we fully expect the "outsourcing" model to play a larger role in voice communication solutions as well.

The great advantage of virtual phone systems is that the only customer premise equipment required is a telephone—and the "customer premise" could be anyplace in the world that has direct dial telephone service. Additionally, these systems do not require a significant up-front investment and the service is usually leased on a month-to-month basis. With minimal setup time and no on-site installation, companies can often start using a virtual phone system in a matter of days, if not hours.

Feature	PC-PBX	Communication Server	IP-PBX
Cabinet and Backplane	Standards-based PC	Proprietary appliance	Proprietary appliance or a packet-switched data network
Control System	PC-based applications running on the PC-PBX platform	PC-based applications running on an external PC server and/or software embedded in the appliance	PC-based applications running on an external PC server and/or software embedded in the appliance
Switching Matrix	Primarily circuit-switched, may include an optional gateway to packet-switched networks	Primarily circuit-switched, may include an optional gateway to packet-switched networks	Primarily packet-switched, may include an optional gateway to circuit-switched networks
Trunk Interface	Analog or digital line cards, may include gateway to packet-switched data network	Analog or digital line cards may include gateway to packet-switched data network	Data network, may include gateway to analog or digital PSTN lines
Station Interface	Analog line cards, or the data network for CTI applications	Analog or digital line cards, or the data network for CTI applications	Data network
Terminals	Analog phones or PC-based CTI applications	Analog, digital phones, or PC-based CTI applications	Analog phones, IP-phones or PC-based CTI applications
Expansion Bus	PC expansion bus, may also use the data network for linking multiple systems	Proprietary bus to connect appliances, may also use the data network for linking multiple systems	Data network
Ancillary Application Servers	Usually not required as all applications run on the PC-PBX platform	Enhanced applications usually run on PC servers	Enhanced applications usually run on PC servers

Table 3-1: Summary of next generation phone system categories.
Note: this is to be used as a general guideline; specific systems will vary.

These services are a great choice for smaller organizations that want to project a big company image, and for geographically distributed companies with multiple sites, mobile workers, or telecommuters. Virtual phone systems are discussed in detail later in the book.

3.4 Migrating the Traditional PBX

No discussion of next generation phone systems would be complete without looking at the reaction of traditional PBX manufacturers. Traditional PBX manufacturers have been providing voice communication solutions to customers for many years. They understand the market and they have mature sales and distribution channels. Although traditional PBX manufacturers have been slower to embrace open, PC-based architectures and data networking protocols, many of these manufacturers have announced strategies or products to compete with next generation solution vendors. After we look at the next generation solutions in more detail, we will explore the migration strategies of traditional PBX manufacturers.

3.5 Summary

To comprehend the variety of solutions on the market and to accurately compare these solutions, we've introduced a categorization scheme based on the primary characteristics of next generation solutions. The primary characteristics that we use for this categorization are:

- Infrastructure requirements and support for convergence
- System integration architecture and interfaces
- Vendor focus

By applying these characteristics to the solutions on the market we see that there are three main categories of solutions that emerge: PC-PBXs, Communication Servers, and IP-PBXs. The chapters that follow are dedicated to exploring these product categories in detail.

Chapter 4

The PC-PBX

4.1 Introduction

As we discussed in Chapter 3, solutions that fall into this category are systems that are integrated into a standard PC and rely primarily on circuit-switched technology for transmitting real-time media. In addition, PC-PBXs typically exploit the programmability of the PC platform to provide a wide range of enhanced applications.

Of course, most of these solutions also support transmitting voice and/or video using packet-switched technology. Usually, this support comes in the form of an optional gateway that allows the system to use a packet-switched data network as a trunk or station interface. However, in general the solutions listed in this category do not rely exclusively upon packet-switched technology for the trunk or station interface. Furthermore, the majority of customers using these solutions today are using only circuit-switched technology for real-time voice communication. Most PC-PBX vendors position their support for packet-switched data networks as a migration path to a converged infrastructure—and a way to compete with IP-PBXs.

One of the earliest PC-PBXs to arrive on the market was the Dash Office Phone System, an MS-DOS based solution introduced in 1993 by Lenexa, Kansas-based Dash, Inc. (Dash, Inc. was acquired by Picazo Communications, Inc. in 1997. Picazo was subsequently acquired by Dialogic, an Intel Company, in 2000). The Dash Office Phone system was one of the first phone systems to include built-in voice mail, auto-attendant, ACD, conferencing, and a Windows-based CTI interface. Although the solution was quite successful—Pizza Hut and Stanford University were amongst the thousands of Dash customers—it was soon eclipsed in the mid-1990s by a wave of vendors who introduced solutions based on Microsoft Windows NT. From the mid-1990s to the present, Windows NT has been the platform of choice for PC-based phone system vendors. Recently, however, a number of companies have announced PC-PBXs based on Linux, a popular open source version of UNIX.

In this chapter, we'll take a detailed look at the PC-PBX segment of the market. In addition to a round-up of products, we'll also look at the key trends and relevant standards that are driving the industry.

4.2 Market Trends

PC-PBXs have the longest history of any of the next generation phone systems discussed in this book. Since the very first PC-PBXs were introduced in the early 1990s, the computing, networking, and telecommunications industries have undergone major changes. These changes have had—and continue to have—a significant impact on the evolution of PC-PBXs. Below is a current snapshot of the major trends that are shaping the market for these solutions.

- Packet-switching technology is playing an increasingly important role in the design and implementation of PC-PBXs. Although it is difficult to foresee any of these solutions relying solely on packet-switched technology for the trunk or station interface in the near future, most of these solutions are beginning to provide migration paths from circuit-switched to packet-switched network integration.
- Standards are emerging that will allow true interoperability of multi-vendor telephony hardware and software in the same PC. In many ways this has always been the holy grail of the PC-based telephony industry. Until recently, competition and turf wars have prevented the industry from reaching a consensus that would allow for such a breakthrough. Today, under the framework of the Enterprise Computer Telephony Forum (ECTF), vendors from across the industry are working together to develop a body of standards that can be used to real-

IP Telephony Gateways

An IP telephony gateway (also referred to in this book simply as a "gateway") is a hardware and software solution for converting real-time media between a circuit-switched network and a packet-switched network. IP telephony gateways were initially deployed by long-distance service providers to extend the cost savings benefit of Internet telephony to anyone with a telephone. Today, many vendors of circuit-switched phone systems offer a gateway for placing or receiving calls on packet-switched networks. The gateway is either integrated in the same chassis with the rest of the system or it is an external, turnkey unit. In either case, the basic function of the gateway is the same - to allow a circuit-switched phone system to take advantage of the efficiency, cost-savings, and features of a packet-switched network when transmitting real-time media.

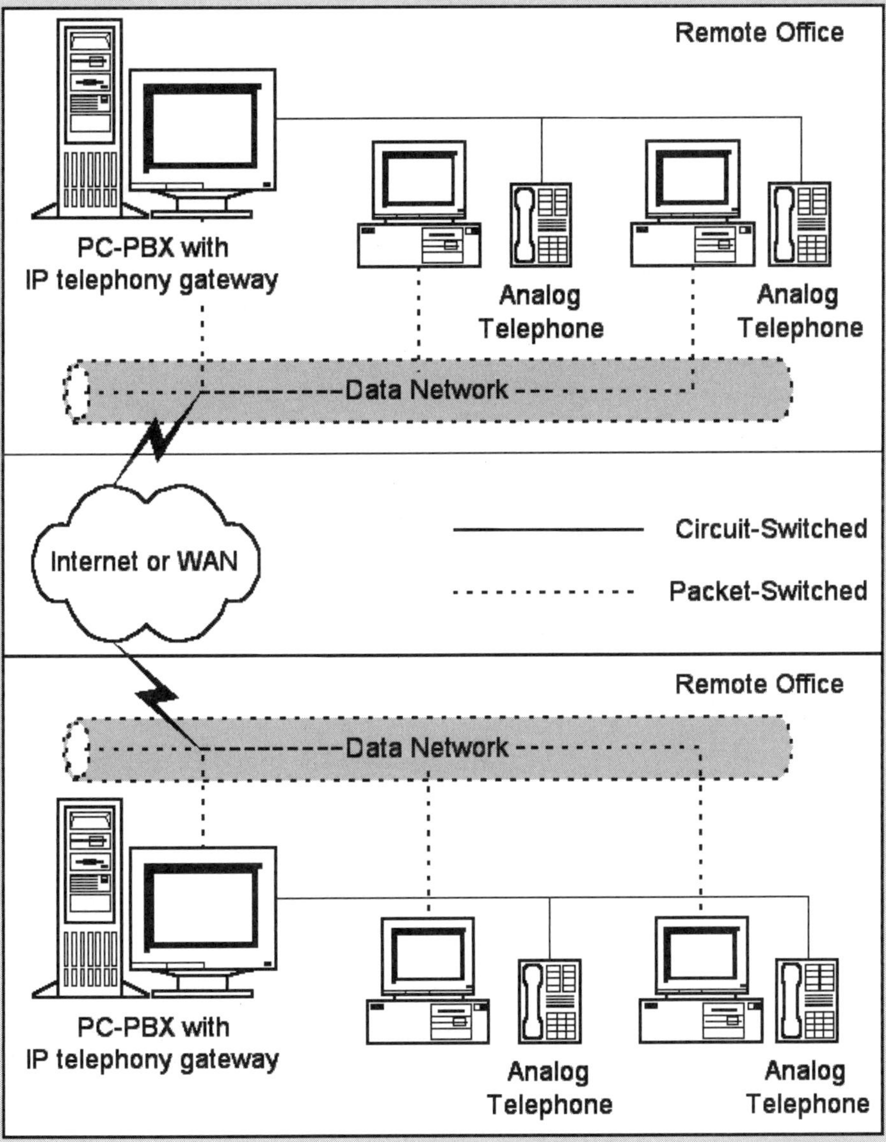

Figure 4-1: Using PC-PBXs with IP telephony gateways as a long-distance toll bypass for connecting the phone systems of remote offices over the Internet for corporate WAN.

ize the vision of a telephony server capable of running applications from multiple software vendors and using telephony expansion cards from multiple hardware vendors.

The work of the ECTF is essentially a reference architecture for developing the components of a PC-based telephony system. Prior to the emergence of the ECTF standards, PC-based telephony applications—although developed on an open, standards-based PC—were proprietary. Customers were typically locked into a vendor-specific telephony hardware and software platform.

Applications developed in compliance with ECTF standards will be able to coexist on the same server, giving consumers the ability to mix and match the specific applications that meet their needs. In addition, the investment that consumers make in a telephony hardware or software platform will have additional protection, since components can be more easily reused in other applications.

One of the first vendors to implement a server platform based on the ECTF reference architecture is Dialogic, an Intel company. Dialogic's CT Media is an ECTF compliant platform for application developers and hardware manufacturers that runs on Windows NT. To better understand the role played by CT Media, we can compare it to the role played by a computer operating system, such as Windows NT. Just as Windows NT makes it possible for applications from different software manufacturers to run on a variety of PC hardware platforms, so does CT Media make it possible to run multi-vendor telephony software applications on an open telephony hardware platform. Just as Windows NT manages and controls a software application's access to PC resources (processor, memory, I/O devices, etc.), CT Media manages and controls a telephony application's access to telephony resources (line interfaces, media processing resources, conferencing resources, etc.).

CT Media is a breakthrough for the PC-based telephony industry. Although few PC-PBXs support CT Media today, most view CT Media as a strategic platform for future development.

- PC-PBXs provide enhanced support for the diverse requirements of workgroups and work-styles throughout an enterprise. Modern corporations have more sophisticated communication requirements than ever before. It's not unusual to find organizations with a large percentage of remote employees, telecommuters, employees who spend most of their time traveling, call centers composed of both on-site and remote workers, plus a variety of other unique work arrangements. Accordingly, the phone system in a modern organization needs to be extremely flexible and configurable.

  With each product release cycle, PC-PBXs continue to meet the demands of the modern organizations. By providing features such as sophisticated call handling rules, call-forwarding, call-screening, full support for remote workers, and Web-based interfaces, PC-PBXs are flexible and powerful enough for almost any organization. As business communication requirements continue to change, PC-based solutions are keeping pace.

- PC-PBXs provide applications to support modern call centers. With most traditional PBXs, call center applications, such as ACD systems, are high-priced add-ons. Most PC-based solutions acknowledge the fact that today almost every organization has some type of call center—even if it is an informal arrangement of employees who handle inbound calls in additional to other responsibilities. Accordingly, basic ACD capabilities are standard with almost all PC-PBXs. In addition, some solutions provide much more sophisticated ACD applications that include the ability to queue and distribute requests in a variety of media formats, including voice, text chat, e-mail, fax, and more.
- PC-PBXs integrate with popular messaging servers to support unified messaging applications. Unified messaging is one of the killer apps of computer telephony. The ability to receive all messages—e-mail, voice, fax, etc.—in a single inbox is a powerful concept. Most PC-PBXs acknowledge the value of unified messaging and integrate with popular messaging back-ends such as Microsoft Exchange.

The Versatile PC-PBX

With industry heavyweights such as Cisco and 3Com promoting pure packet-switched voice solutions, many wonder what the future holds for the PC-PBX. The PC-PBX of today is designed around circuit-switched voice technology and many industry observers predict that such "legacy" technology will be replaced by packet-switched solutions. Although it is agreed that packet-switched solutions will become increasingly prevalent, there isn't any sign that the PC-PBX will disappear. In fact, it seems that the PC-PBX will thrive in a world of packet-switched voice solutions.

Why is that? It is because the PC-PBX is the most versatile voice communication solution on the market. The programmability of the PC-PBX guarantees its survival, regardless of the technology trends of the future. Just as PC-PBX manufacturers have added VoIP gateways and many other enhancements to their platforms, so too will they support the standard call control and media protocols of packet-switched voice communications; many already do. In a packet-switch voice future, the PC-PBX will be one of many intelligent voice communication devices connected to the network, providing enhanced applications such as ACD, voice mail, auto-attendant, circuit-switched network integration, and more.

4.3 Relevant Standards

In Chapter 1, we mentioned that PBX functionality can be divided into two main categories: call control and media processing. A PBX is essentially a device that provides real-time control of a media stream. Accordingly, to achieve system interoperability, standards are required for both the definition and control of the media stream.

4.3.1 The Enterprise Computer Telephony Forum

One of the most active standards organizations addressing the requirements of PC-based telephony is the ECTF. The ECTF is composed of hardware manufacturers, software manufacturers, and service organizations with an interest in developing the computer telephony industry. ECTF membership includes Lucent, Nortel Networks, Mitel, Siemens, Dialogic, Brooktrout, Microsoft, Sun, and many other companies. To date, most of the ECTF's work has concerned interoperability of applications and components on the PC server platform. Recently, however, the ECTF has extended the scope of their framework to include IP networks, wireless, application scalability, and multimedia (see the *Insider Perspective* at the end of this section).

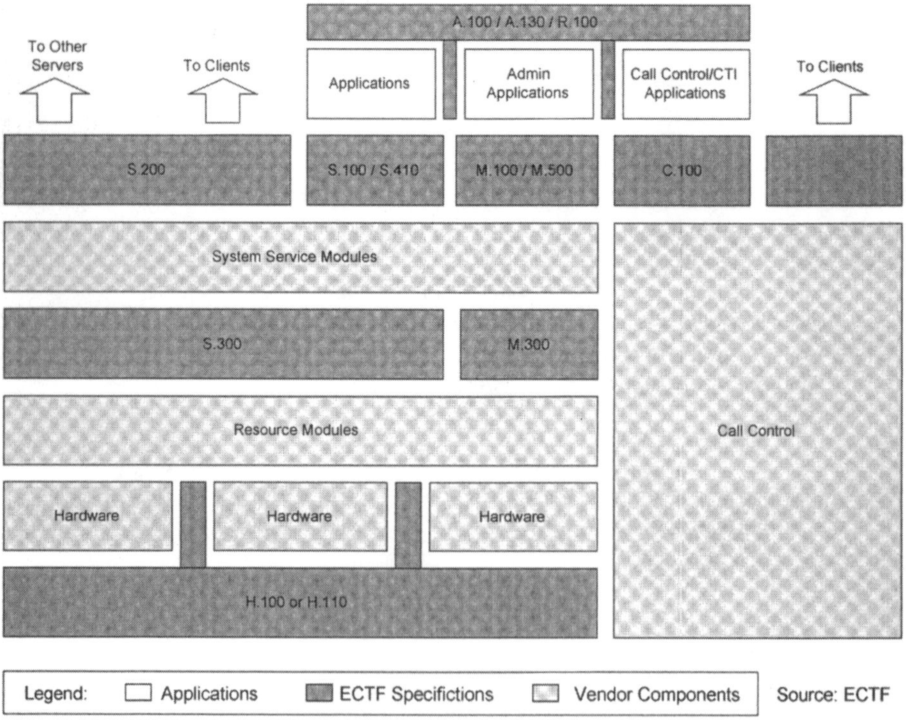

Figure 4-2: The ECTF framework for an open, PC-based telephony server.

The ECTF has published a framework that defines the components and interfaces of an open, PC-based telephony platform (see Figure 4-2). Within this framework, there are specific working groups that define and develop the standard interfaces.

An overview of the most significant and well-defined interfaces in this framework is provided in the chart below.

ETCF Standard	Description
A.100 and A.130	ECTF A.100 and A.130 are standards that describe the interoperability of S.100 compliant applications. A.100 describes call handling between applications and A.130 describes data sharing.
S.100 and S.300	S.100 and S.300 are media processing specifications for PC-based telephony servers. S.100 is an interface standard that allows application programs to allocate, configure, and operate media processing resources. S.100 is designed to be both operating system and hardware vendor independent. S.300 is the specification that describes a standardized interface to hardware resources and provides hardware-vendor interoperability.
C.100	C.100 is a specification for the interoperability of call control services in computer telephony applications. ECTF C.100 specifically refers to an agreement that certifies Java Telephony API (JTAPI) 1.3 as ECTF compliant. The ECTF has now become one of the lead organizations in the future development and evolution of JTAPI.
H.100 and H.110	H.100 is a specification that describes the interoperability between computer telephony hardware components in a PC-based server. H.100 describes the physical interface of the bus used to link PCI form-factor computer telephony resource cards. H.110 is an addendum for CompactPCI form-factor cards. The introduction of H.100/H.110 brought an end to the competition between the incompatible Signal Computing System Architecture (SCSA) bus and the Multi-Vendor Integration Protocol (MVIP) bus.

Table 4-1: Summary of ECTF standards.

PC Telephony Expansion Buses

Just as the ISA and PCI expansion buses are integral components of a PC, telephony expansion buses—designed to carry real-time audio between telephony expansion cards—are an important part of PC-based telephony servers. The PC-based telephony expansion bus is implemented as a multiple-drop ribbon cable that interconnects a variety of telephony hardware, including telephone network interface cards, voice processing cards, speech recognition cards, speech synthesizers, multi-line fax cards, etc. within a single server.

Why not just push audio through an ISA or PCI bus instead of creating yet another expansion bus? Well, the hardware that supports telephony applications must meet special requirements, such as support for a precisely timed 64Kbps data stream. Telephone networks are extremely intolerant of devices whose audio sample rate drifts away from the sample rate used in the public network. Thus, it is important to push audio data through a bus whose data rate is precisely timed, and can also be slaved to an outside synchronization signal (which is often the case if the server is connected to the PSTN via digital phone lines).

ECTF H.100 and H.110 address standardization of the PC telephony expansion bus and bring an end to the "bus wars" between supporters of the incompatible SCSA and MVIP telephony expansion buses.

Bus Attribute	**SCSA**	**MVIP**	**H.100/H.110**
Number of Time Slots	2,048	512	4,096
Data Rate	8 Mbps	2 Mbps	8 Mbps
Number of Data Streams	16	16	32

4.3.2 Microsoft TAPI

Another major interface standard influencing the evolution of PC-PBXs is Microsoft TAPI (Telephony Application Programming Interface). TAPI is a Microsoft API that is part of the Windows system architecture.

Microsoft delivered the first version of TAPI on the Windows 3.1 platform, and subsequently on Windows 95. TAPI 1.0 was designed as an API to provide first-party call control. The primary use of TAPI 1.0 was to allow software, such as a contract manager or PIM, to initiate or answer phone calls via a standard PC modem.

TAPI 2.0 was designed to run on Windows NT and provide third-party call control. TAPI 2.0 is used by application programs to manage and control telephony resources within a Windows NT-based communications server or in an external telephony platform, such as a traditional PBX. Telephony vendors provided support for TAPI by developing a specialized driver (called a service provider) that provides the interface between TAPI and their platform—even if that platform is external and connected to the Windows NT server via a communications link.

First-Party vs. Third-Party Call Control

Call control functionality can be broken down into two types: first-party call control and third-party call control. First-party call control can be compared to what you as a telephone user do whenever you place or answer a call. You are a participant, or first-party, in the call that you are acting upon. Third-party call control can be compared to the behavior of an office switchboard operator who connects and controls the call of other parties or users of the system. Call control standards address both first-party and third-party call control.

Microsoft released TAPI 2.1 to bring TAPI 2.0 functionality to Windows client PCs on a network. TAPI 2.1 is used for client/server telephony applications. Through the release of TAPI 2.1, Microsoft's focus was to provide an interface for telephony call control, not media processing. This, however, changed with the release of TAPI 3.0.

TAPI 3.0 was delivered with Windows 2000 and incorporates both call control and media processing APIs. TAPI 3.0 provides a number of significant enhancements, including a simplified programming interface based on COM (now TAPI applications can be written in any language, not just Microsoft Visual C/C++ or Microsoft Visual Basic), support for IP telephony, integration with Microsoft's Active Directory (or any LDAP directory), and support for low-latency streaming of media through Windows driver model streaming architecture (see Windows 2000 and Telephony, below).

One of the design goals for TAPI 3.0 was to provide a single interface that integrates PSTN telephony with IP telephony, allowing telephony applications to be built using a single API regardless of the underlying network.

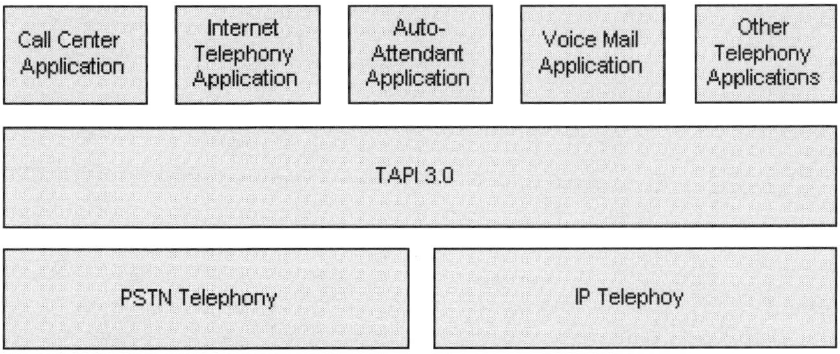

Figure 4-3: TAPI 3.0 integrated PSTN and IP telephony, allowing telephony applications to be written to a single interface.

4.3.2.1 TAPI and the ECTF

Historically, the ECTF S.100 and ECTF S.300 standards have been complimentary to Microsoft TAPI—the ECTF standards primarily described media processing and TAPI primarily described call control. However, with the release of TAPI 3.0 and ECTF C.100, the ECTF framework and TAPI begin to overlap.

TAPI now includes high-performance media processing and the ECTF C.100 (JTAPI) agreement includes call control. When and if TAPI will be completely merged into the ECTF framework—or vice versa—remains an unanswered question. However, with Microsoft's involvement in the ECFT, we would hope to see the convergence of these two telephony frameworks sometime in the not too distant future.

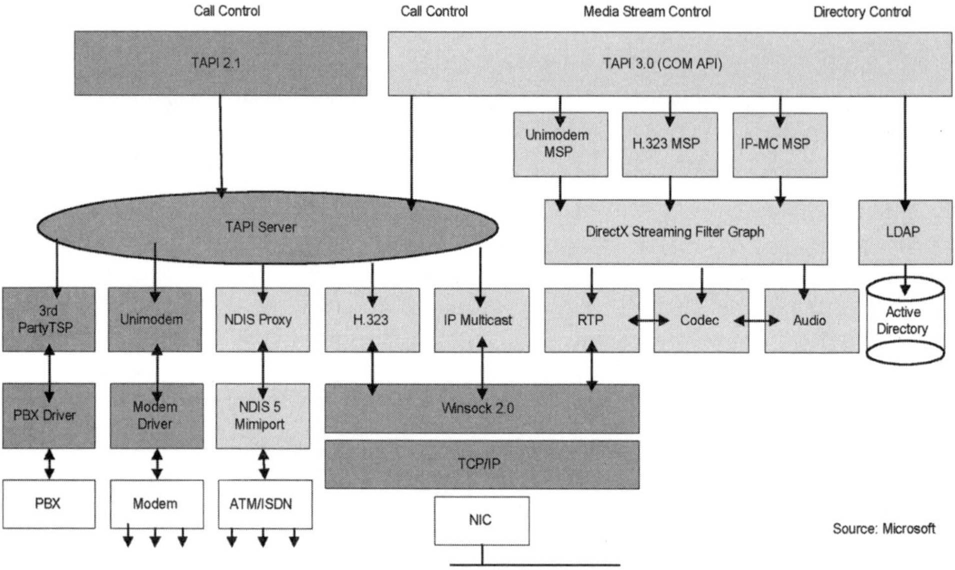

Figure 4-4: Microsoft TAPI architecture.

4.3.3 ITU H.323

In addition to the work of the ECTF and Microsoft, there are a number of standards from the International Telecommunications Union (ITU) that are affecting the evolution of PC-PBXs. The most important of these standards is H.323. ITU H.323 defines voice, data, and video communication over packet-switched networks, including the Internet. H.323 provides specifications for call establishment and teardown, call control, capability exchange, supplemental services, and more. At the present time there is minimal interoperability between H.323 implementations from different vendors. This is mainly because the definition of H.323 version 1 was not rigorous enough to ensure interoperability. H.323 version 2 does a much better job of addressing interoperability.

PC-PBX vendors are using H.323 as part of a distributed architecture that allows multiple PC-PBXs to be connected across a LAN, WAN, or the Internet. Accordingly, H.323 is a core technology for building distributed PC-based phone systems. Additionally, a number of vendors have announced support for H.323 as a means of supporting multi-media desktop applications such as Microsoft Windows NetMeeting or IP desktop telephones.

As PC-PBXs integrate more tightly with packet-switched networks, standards like H.323 are becoming increasingly important.

Windows 2000, TAPI 3.0, and Telephony: An Interview with Microsoft and Dialogic

Windows 2000 is Microsoft's latest incarnation of their business-class operating system. Win2K merges the easy Win98 interface with the strong foundation of WinNT 4.0. While Win2K is sometimes portrayed as the next OS for both Win98 and WinNT 4.0 users, it is clearly an upgrade of WinNT 4.0 and is targeted for business users.

As telecoms professionals, we may be wondering what does Win2K have in store for telephony? As one who does not like unanswered questions, I set out to see what the future of telephony would be like on the new Win2K platform. Read on to see what I found out.

Win2k

Win2K was developed to follow a few trends in the industry. First, there was a push for convergent solutions that brought together aspects of telephony, electronic messaging, PC-based applications, and IP-based applications. Second, there was a clear desire to keep the underlying platform open, which in Microsoft parlance means that anyone can write for the platform (not that the OS is open source). This open platform can then be used to take advantage of the deregulation and convergence of emerging communication platforms. Finally, the ways in which people communicate have evolved to place more emphasis on message-rich customer contact centers and data-intensive mobile platforms.

The main Microsoft platform for the development of communication solutions has been WinNT 4.0. Microsoft understood that to remain integral to future communication solutions, it needed to enhance the platform—hence Win2K.

I spoke with Mark Lee, Microsoft's Product Manager for Windows Telephony, and with Jason Bailis, from Dialogic. They clued me in on the telephony improvements found in Win2K. For some reason, I had expected the bulk of the telephony improvements in Win2K to be in TAPI 3.0. But, it turned out that the bulk of the improvements were in the reliability, manageability, scalability, and device support of Win2K. Let me briefly touch on these improved features.

Win2K Reliability

There have been numerous improvements to reliability with the chief benefit being increased uptime. For example, improvements in managing protected memory, so that one little application doesn't bring down the whole system. Also, Microsoft has reduced the number of conditions that require a system reboot, so that during an installation or a crash, the system stays available. And in multi-server environments, automatic network load balancing allows for a system to be dynamically removed from and restored to the network without affecting the rest of the network. This improves the reliability and availability of the system as a whole. Moreover, Microsoft has designed a logo certification program that verifies a product's performance, interoperability, and functionality within Win2K.

Win2K Manageability

Microsoft has developed an enhanced set of manageability features into Win2K. For example, there is a management console that allows administrators to view the state of a system. Central to Win2K system management is the Active Directory, which provides access to all the network and system properties. It provides a central and extensible method of resource administration. And because it is extensible, other developers can improve or add to the Active Directory to enhance analysis and management of current or new resources.

Mr. Lee further pointed out that Win2K understands the different Quality of Service (QoS) requirements for different networks, such as switched or routed. In addition, Win2K has native support for ATM (in WinNT 4.0 it was more of an abstraction) and allows for a more granular control of QoS over ATM. These QoS issues stand to make Win2K a good citizen in the telcom environment.

Win2K Manageability—Device Support

Within this manageability, Win2K also provides for Plug-and-Play support for new devices, making installation and support easier. Also, Win2K can offload processing to smart devices to increase CPU availability. Mr. Lee pointed out that this is a big issue with TCP/IP and some other routing protocols.

Win2K also allows for a better degree of scalability, allowing more servers per computer (up to 32) and also dynamic load balancing to share processing among a group of servers.

Overall manageability has improved versus WinNT 4.0 and is well suited for parallel systems that need high availability, such as 'Net or telecom servers.

Win2K Price/Performance

Mr. Lee pointed out some serious changes in the performance of Win2K relative to WinNT 4.0. He told me that the combination of WinNT 4.0 and SQL Server placed first in the price/performance ratio, relative to similar offerings from other platforms. But, when looking at performance alone, WinNT 4.0 and SQL Server dropped far down the list. Mr. Lee was happy to tell me that Win2K and SQL Server maintained the lead in price/performance, and has become number one in performance. In addition, the price is almost 40% less, relative to the next best offering. So, not only is Win2K/SQL Server the price/performance leader, but according to Mr. Lee, it is number one in price *and* performance.

Win2K—TAPI 3.0

As the core telephony API, TAPI 3.0 has been improved over TAPI 2.1. For example, because the API is COM-based, programs can be written in any language, not just Visual C++. Various IP telephony components are now native to TAPI 3.0, allowing easier development of VoIP systems. Call control and media services, such as media streaming are modeled as telephony or media service providers (TSPs or MSPs). SPs are responsible for resolving the protocol-independent call model of TAPI into protocol-specific call-control mechanisms. MSPs implement Microsoft Direct Show interfaces for a particular TSP and are required for any telephony service that makes use of Direct Show streaming. Additionally, TAPI provides for IP Multicasting. Existing TAPI 2.1 programs will continue to function in Win2K, but, due to the underlying changes (COM-based), programs need to be completely rewritten to take advantage of the new features.

Microsoft has included in Win2K an updated Phone Dialer that takes advantage of the new TAPI 3.0 functions. Mr. Lee mentioned that the new version of the Phone Dialer was written by Active Voice. The Dialer supports IP Multi-casting and H.323v1 (TAPI 3.1 will support H.323v2). The version of NetMeeting that comes in Win2K is not TAPI 3.0-based. It does not support multicasting, but can support H.323v2. Mr. Lee expects the forthcoming TAPI 3.1 to support H.323v2.

IWin2K CT and IP Telephony

For me, the most interesting development in Win2K is that Microsoft put more IP-based functionality into the system. A gem hidden in the platform SDK is the Web Telephony Engine. This Web server add-on provides the tools to build telephony applications using HTML tags to hook into TAPI and

SAPI. Web sites easily can be made available for telephony-based access. In short, developers can merge 'Net-based applications with telephony applications, such as IVR, voice mail, or ACDs. Mr. Lee mentioned to me that the Engine was made available after the Win2K function list was frozen; therefore, it's presence in the SDK and not the general release. Incidentally, the Engine was developed primarily in Israel.

Win2K Telephony Resources

I asked Mr. Lee about resources Microsoft provides for Win2K Telephony developers. He mentioned the Hardware/Software Developers Programs and developer communications resources. He also mentioned that there are 'relation groups' that work with specific vendors, such as ActiveVoice, Artisoft, and Dialogic. Part of this relationship is the logo program that is not just a compatibility stamp, but requires that the product use Win2K features, such as the Active Directory or TAPI 3.0. For example, Pivotal is an eCRM software developer that has the Win2K logo. ActiveVoice's fax software also carries the Win2K logo. These products not only work in Win2K, but use manageability and other Win2K-specific functions. Mr. Lee pointed out that Cisco had recently announced a VoIP phone system that has a Win2K core. While the product uses many Win2K-specific features, he is not sure if it has received the logo, yet. In summary, Mr. Lee said that this logo 'raises the bar' on the quality level of Win2K products.

Win2K Dialogic

I spoke with Dialogic, because of its central role in CT. Mr. Bailis mentioned that all Dialogic products were Win2K compatible a week before the Win2K release. He said that Dialogic's intent is to allow users to take immediate advantage of the new platform using old software. Users will then be able to evolve their products to take advantage of Win2K-specific features. I was surprised that Mr. Bailis seemed more pleased with the improvements in reliability, scalability, and management, than any changes made in TAPI 3.0. Now that I understand the changes between WinNT 4.0 and Win2K, it is clear to me why.

Win2K Future

I asked Mr. Bailis and Mr. Lee about the conversion rate to Win2K and the phase out of WinNT 4.0 products. Because Dialogic installations are deployed by third-party vendors, Mr. Bailis couldn't give me an exact figure on the conversion rate. But, based on downloads of the software, there seems to be great interest. Mr. Bailis expects Dialogic to keep developing Win2K products in parallel with WinNT 4.0 products over the course of the next year as the installed base migrates to Win2K. Over the same time, Dialogic products will take on more Win2K functions. Dialogic is most interested in providing a smooth transition for its entire installed base.

Mr. Lee said that Microsoft would still be offering WinNT 4.0 for some time, but that new installations would most likely be Win2K. Also, there is a significant installed base of WinNT embedded devices. Embedded WinNT is a component version of WinNT 4.0 that allows developers to install only necessary components for a smaller, faster, and more stable WinNT 4.0 solution. Mr. Lee didn't expect embedded WinNT 4.0 to be replaced any time soon. Many of the vendors who use embedded WinNT 4.0 develop hardware platforms that do not change as readily as open consumer platforms and still have a lifetime of many years. Yet, I am sure that an embedded version of Win2K is in the works. In summary, WinNT 4.0 will be available for the near future so as to provide stability for current installed systems and solutions and at the same time there will be a push for new systems to use Win2K.

Win2K Take Home Message

What is a take home message on how Win2K is transforming telephony?

- Win2k is more reliable. From protected memory, to load balancing for server farms, to greater system stability, Win2K is designed for systems that need more uptime.
- Win2K is more manageable. At the heart of its manageability is the Active Directory, which allows administrators to peer deep into the system and network properties. Also, the Active Directory allows for better control of Quality of Service issues for various network standards (especially ATM). Win2K also make it easier to support and use devices through Plug and Play and processor offloading.
- Win2K is more scalable. It provides for up-scaling—up to 32 processors in a system; and out-scaling—software scaling and load balancing for server farms.
- Win2K is more IP friendly. It has made native to the system more network-based standard, such as XML and ATM. TAPI 3.0 provides for better convergence between traditional PSTN-based telephony and IP-based telephony. And, the Web Telephony Engine is a clever add-on that allows for the easy deployment of HTML-based telephony applications.

I think Win2K has promise as a telephony platform. The migration from WinNT 4.0 to Win2K may not be as painful as one might expect—many vendors are already doing the work. Also, the features listed above have increased the robustness and friendliness of the system. As the larger telephony application providers update their offerings, Win2K will become as prevalent a telecom platform as WinNT 4.0 is now. Go out and kick the tires. You may be impressed.

By Charlie Schick

INSIDER PERSPECTIVE

The Push for Interoperability Among Converging Computer Telephony Technologies

By Andrew Hunkins—President and Chairman, ECTF

The ECTF gears up to engineer interoperability across all the standards and specifications of today's converging technologies

Today, a large number of concerned people (developers, hardware designers, system integrators, VARs, distributors, carriers, network planners, system buyers, network administrators, and IT support staff) are all struggling with how to address the many new technologies entering the CT market, each with its own set of complex, incompatible standards and specifications.

Each year finds the CT market becoming a little more complex as it embraces new technologies driven by the Internet engine, which generates increasing demand among consumers for more and more sophisticated applications — applications that go far beyond just data. This growing consumer demand includes applications such as video conferencing and even sophisticated CT applications that can run on their PDAs. It is forcing CT developers and others to address the integration of next-generation networks, IP and wireless transports, and multimedia.

Standards and specifications in each technology have provided a common "language" for the design work, and these standards have streamlined the design process to eliminate most of the drudgery of dealing with compatibility issues. Well-written standards or specifications make it easier to bring mature applications to the market quickly as well as to port those applications to different platforms and environments. The problem then becomes simply one of choosing the right standard.

However, when multiple technologies converge, their existing standards don't provide all the help that is really needed. Developers face big gaps between the standards and specifications of the various technologies.

Providing an Interoperability Framework

Today there is a pressing need to skillfully engineer interoperability between all the different standards and specifications of technologies that are beginning to converge. It's a much more difficult task than addressing compatibility issues within an individual technology. Although unrewarding and time consuming, developers could deal individually with compatibility issues within a technology and still bring products to market in a timely enough fashion to meet consumer demand. However, achieving the same level of interoperability between differing technologies threatens to slow the design process down to a point that will make it very difficult for the integrated technology to meet market demand.

Most specifications were developed independently with no eye to subsequent interoperability, and that often created gaps or overlaps between them that added to the complexities of converging different technologies. Since it is unrealistic to expect developers to address these complex issues on an individual basis, the ECTF is expanding its charter to allow it do the needed work.

The ECTF is well suited to the task of engineering interoperability on a large scale because it has accumulated a considerable amount of experience in defining interoperability for the complex Computer Telephony (CT) market. Interoperability Agreements developed by the ECTF for that market were specifically designed to address the complex issues of CT interoperability. In fact, the ECTF specifically designed its CT Framework to allow for the interoperability of converging technologies.

The ECTF's work in developing interoperability specifications for CT technology has allowed it to accumulate some of the brightest minds in the industry. Aware of its high level of expertise in dealing with interoperability issues, and anticipating the larger interoperability needs of the converging technologies, the ECTF announced its intentions to become a clearinghouse for technology specifications that are converging into the CT sector. By directly addressing interoperability issues between technologies, the ECTF will become a major interoperability resource point for developers to determine the best standards or specifications to adopt.

Many organizations now provide "point" solutions for interoperability in various technologies, but the ECTF will provide a vital holistic perspective to achieve full interoperability between them all. As a consortium of many CT companies working together to engineer interoperability, the ECTF has already proven that it can effectively create the highest levels of interoperability in an environment of divergent technologies. This is clearly exhibited in the ECTF's definition of an effective CT

Framework supported by clearly defined Interoperability Agreements that are available to anyone in the industry.

Unifying the entire field of computer telephony by delivering a complete architecture will involve bringing new abstractions, protocols, and APIs into the ECTF Framework. The ECTF's existing specifications have already helped to generate a first wave of modular CT products. Some companies are already using ECTF-based solutions in public network applications and IP telephone systems, and the ECTF is in the process of expanding its framework by adopting and incorporating other standards. They are also augmenting existing specifications where necessary. These efforts are requiring the ECTF to work closely with organizations that are developing related standards and specifications—a process that has been ongoing for some time.

The ECTF plans to publish Implementation Notes to describe how the new and existing ECTF specifications will apply to converging technologies. These notes will keep developers fully informed about how to accomplish the needed integration. An example of this is the TDM specification for switching fabrics that will be expanded to include an IP switching fabric. The ECTF will fill all of the interoperability gaps between converging technology standards without conflicting with the existing specifications.

The ECTF is uniquely qualified to accomplish this task because it has always represented the full spectrum of the CT market and encourages everyone involved to join the ECTF and become an active member. This openness has attracted the industry's most innovative thinkers with many ECTF members currently serving on other standards bodies. The ECTF's charter mandates that it serve all participants to create a vital market, and the organization has purposely avoided becoming dominated by any one group. It operates as a full consensus organization that provides a variety of membership options to allow any interested party to participate.

It is a good time for the Forum to address trans-technology interoperability because it has completed the majority of its work on the ECTF framework. The few specifications that haven't been introduced yet are mostly complete, which has freed up many ECTF technical resources to address wider interoperability issues. The interoperability integration resulting from the ECTF's activities will enable developers to reduce time-to-market and allow CT technology to keep pace with the demand for new applications.

Getting the ECTF to Review Standards

The ECTF will review any standard it receives for its integration potential. ECTF members can submit a standard to the ECTF for review, and non-member organizations can lobby ECTF members to make the contribution on their behalf. Of course, the best way to contribute is to join the ECTF and directly participate in working group activities. For additional information on the ECTF and to obtain ECTF documents, please access the ECTF web site at http://www.ectf.org. To inquire about membership, contact the ECTF at 510-608-5915 or ectf@ectf.org.

This Insider Perspective was contributed by Andrew Hunkins—President and Chairman, ECTF. Our sincere thanks to Sabrina Ritchie, Marketing Project Manager for the ECTF for her help in coordinating this article.

Chapter 5

PC-PBX Market Overview

This chapter provides an overview of the leading PC-PBXs on the market today. These solutions include AltiGen AltiServ, Artisoft TeleVantage, COM2001.com InternetPBX, Interactive Intelligence Enterprise Interaction Center (EIC), and InVADE International.

5.1 AltiGen AlitServ

AltiGen Communications is one of the trailblazers in the PC-based phone system market. In 1996, AltiGen introduced AltiServ, one of the first Windows NT-based phone systems with built-in auto-attendant, voice mail, ACD, and more. Then in 1998, AltiGen released AltiWare Open Edition, an impressive software upgrade that added features such as Java CTI, TAPI support, an operator console, and messaging platform integration. More recently, AltiGen released a hardware and software IP telephony gateway, which is completely integrated with their PC-PBX.

5.1.1 Market Focus

AltiGen has designed the AltiServ system with two important customers in mind—small and medium enterprises (SMEs) of up to 160 users and departmental call centers of larger organizations.

As business communication evolves, SMEs (which account for over 90% of all businesses) must communicate using an increasing number of mediums, including voice, fax, e-mail, video, and the Internet. In this market segment, AltiGen AltiServ is positioned as a building block around which these communication services can be consolidated. For example, if a business had use AltiGen AltiServ in 1997, it could have ran its phone system, Web site, and Internet mail server from a single Windows NT server.

In the call center market, the ACD and CTI features of AltiServ make it especially attractive. By their very nature, PC-based systems such as AltiServ, significantly decrease the amount of "integration" required to implement CTI features. AltiGen, for example, provides a suite of Java applets that are installed on the server and provide CTI functionality to any user through a Web browser—no need to install a costly CTI link or configure an "integration" server.

5.1.2 Features

At the heart of an AltiServ system is AltiGen's own computer telephony hardware platform known as the Quantum board. The Quantum board has been specifically designed as an all-in-one platform that includes an analog network interface, voice processing resources, a telephony power supply, and support for standard analog extensions.

Each Quantum board supports 12 ports that can be allotted between outside trunk lines and internal extensions. With AltiGen's digital T-1 network interface cards, an AltiServ system is scalable up to 160 users—enough for most SMEs and departmental call centers.

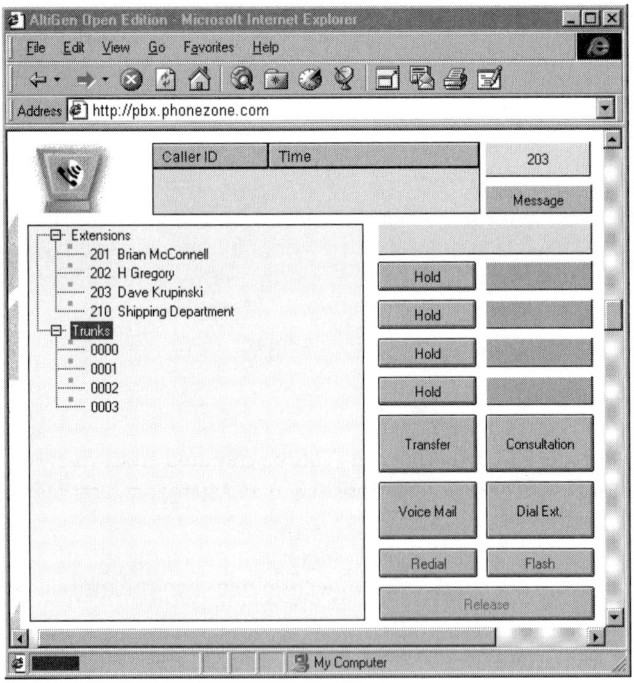

Figure 5-1: AltiGen's Java CTI application runs in a web browser.
Copyright (c) AltiGen Communications, Inc.

AltiGen's AltiWare Open Edition software is tightly integrated with the Quantum platform and provides a range of features, including:

- Programmable auto-attendants.
- Voice mail—Voice mail includes a feature known as "boomerang" that allows users to return calls in the course of listening to their voice messages.
- Automated Call Distribution (ACD)—The ACD system supports up to 32 work-groups with 64 members per group, including the ability to have users in more than one ACD workgroup. A Java applet allows users to view and manage the call queue from a Web browser.
- Integrated CTI—Users can perform most call control activities—dial, hold, transfer, forward—and see the status of other users right from their PC screen.
- Telecommuter features—AltiServ provides comprehensive follow-me roaming support and the ability to change personal configuration settings through a Web browser.
- Standard analog phone support—AltiServ will send caller identification information to standard analog extensions.
- Third-party application support—AltiServ's open architecture includes full TAPI 2.1 support to run third-party CTI applications.
- Backup and restore software—AltiServ automatically copies all configuration files and recorded voice prompts to a backup location at regularly scheduled intervals.
- Much more—AltiServ also includes a POP3/SMTP Internet e-mail server, Microsoft Exchange integration, an optional Operator Console application for receptionists, and music-on-hold support.

In addition to AltiWare Open Edition and the Quantum platform, AltiGen also provides a hardware and software bundle for IP telephony support. AltiWare IP is an H.323 compliant IP telephony gateway that supports both PC-to-phone and phone-to-phone calls. While PC-to-phone calls can be initiated from many popular H.323-compliant desktop software applications (such as Microsoft NetMeeting), phone-to-phone calls are limited to calls between AltiGen AltiServ systems at the present time. With AltiWare IP, AltiGen is targeting companies that have offices in multiple locations with an existing data network connection between the offices. For such organizations, AltiWare IP can be used for long distance toll bypass while providing a migration path for merging voice and data networks.

AltiWare IP is designed around a hardware expansion card that AltiGen has dubbed the Triton board. The Triton board is a full-size PCI bus expansion card for Windows

NT. Each DSP-based Triton board can support up to 4 IP telephony sessions. AltiGen has designed a virtually seamless system that allows IP telephony calls to take full advantage of the features built into the AltiGen platform, including auto-attendant, voice mail, ACD, and more. Additionally, PSTN and IP calls can be conferenced together and calls can be transferred between PSTN and IP trunks.

IP telephony trunks are configured the same way that PSTN trunks are setup. The "Numbering Plan" administrative screen is used to identify which digit selects an IP trunk. For example, you might configure the system to use "9" to select a PSTN trunk and "8" to select an IP trunk (see Figure 5-2). Additionally, a dialing plan can be established to associate a short string of digits with an IP address. If the digits "01" are associated with the IP address of a remote office, then dialing "801" will connect the caller with the AltiGen system at the remote office.

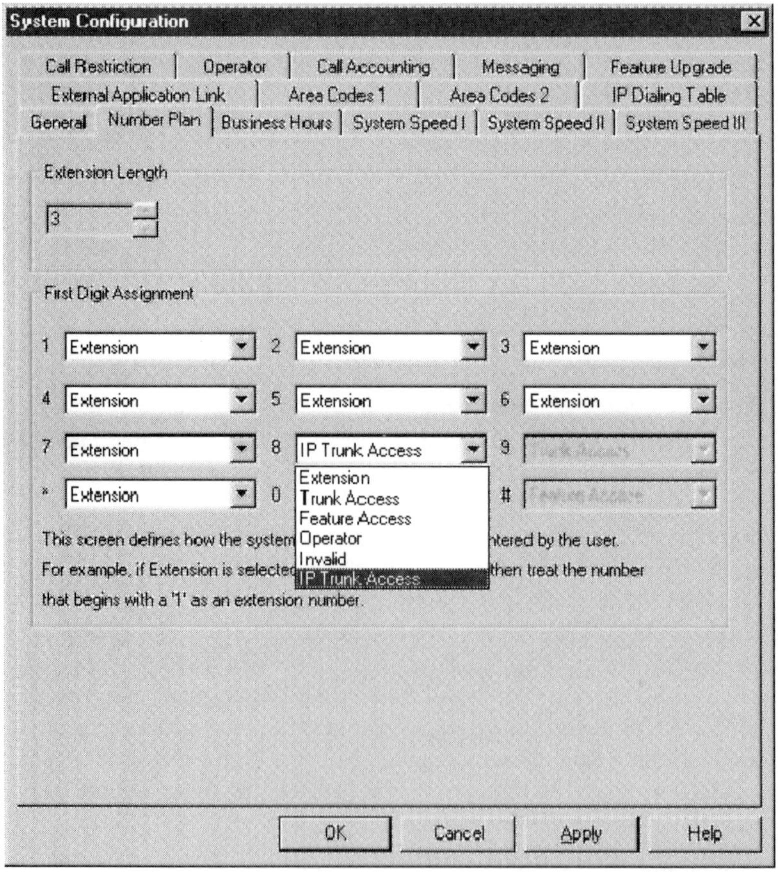

Figure 5-2: Configuring AltiWare to access an IP telephony trunk when a dialed number begins with "8". Copyright (c) AltiGen Communications, Inc.

AltiWare IP supports G.711 (64 Kbps) and G.723.1 (5.3 and 6.3 Kbps) voice compression. AltiWare IP also provides a configurable jitter buffer to accommodate for the throughput variability of IP networks.

5.1.3 Test Drive

For my testing, you could install a single AltiGen Quantum board and AltiGen AltiWare Open Edition software. The Quantum board should be configured to support 4 analog trunk lines and 8 extensions. There should be no dip switches or jumpers that need to be set on the board. The only hardware configuration required would be setting the board id (0 to 9) with a dial on the rear plate of the board. The first board in a system is given an id of 0, the second board is given an id of 1, and so on.

Then insert the board into a full-length ISA slot on a 200MHz Pentium with 32 MB of RAM running Windows NT 4.0. Although the current revision of the board, revision D, does not require an external power supply to ring the extensions, this test uses a revision C board, which does require the power supply. The power supply connects to the Quantum board via a cable plugged into the board and connected to a rear plate of a vacant PC expansion slot.

Before turning on the PC, connect a harmonica style patch panel to the 25-pin connector on the Quantum card. The patch panel will have 12 RJ-11 jacks—the first four jacks are used for the trunk lines and the remaining 8 are used for the extensions. Now turn on the PC, insert the Open Edition CD and install the software.

Upon starting the AltiWare Open Edition software presents the main graphical administration screen. This screen displays the status of the trunks and the extensions and also shows a log of all inbound and outbound calls. Using the point-and-click menus, complete the setup by creating a simple auto-attendant, configuring the extensions, and specifying some general information, such as the local area code and business hours.

With AltiWare Open Edition, the administration program can be run from any workstation connected to the company network, making it even easier to maintain and update the system

For the average user, the AltiGen system operates just like any other phone system. It supports all of the basic features that users expect, such as voice mail, do not disturb, and conventional hook-flash based transfers and conference calls. However, for the growing number of organizations that demand a more flexible communications platform to support workgroups, remote workers, traveling employees, or employees who spend a significant amount of time on the phone, AltiServ really shines.

One of the authors favorite AltiServ features is the intelligent manner in which it uses caller identification information. AltiServ stores caller identification information with each voice mail message, allowing return calls to be placed while retrieving messages without needing to dial the number—AltiServ does it for you. After placing the call, you are returned to your voice mailbox where you can continue to listen to messages. This is an especially useful feature when you are out of the office, since you only need to place one call to your AltiServ system to retrieve and respond to all of your messages. AltiGen takes this one step further by providing a call-bridging feature that allows you to dial any number as if you were in the office, even though you might be calling from a remote location.

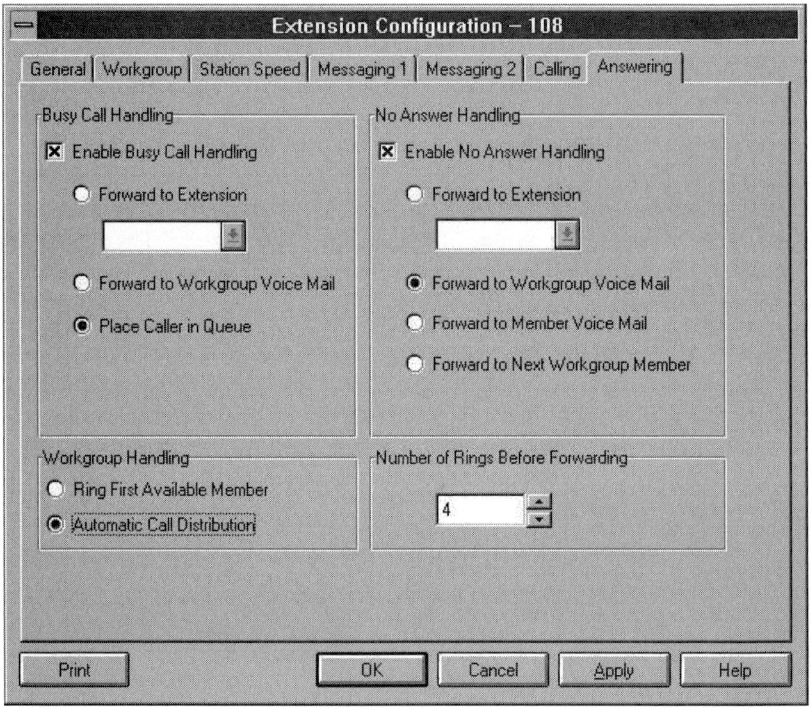

Figure 5-3: Extensions are configured with a graphical user interface.
Copyright (c) AltiGen Communications, Inc.

Another exciting feature is AltiServ's support for telecommuters and remote workers. AltiServ allows you to configure a detailed call-forwarding scheme to locate you when you are out of the office. This call-forwarding scheme can be updated from any Web browser. For remote power users, AltiServ offers additional features. For example, any user can be configured with a personal ACD to queue incoming calls. The queue can be managed and viewed from any Web browser, regardless of whether you are in the office or working from a remote location.

AltiServ's workgroup features are also impressive. Users can dynamically log in and out of workgroups and AltiServ will automatically distribute calls to the members of the workgroup. If all workgroup members are on the phone, incoming calls will be queued and callers will periodically be given the option to leave a message or continue to hold. Coupled with the Java CTI applets, the ACD features are sufficient for most organizations.

5.1.4 Summary

AltiGen AltiServ is a rare combination of ease-of-use and a complete, well-conceived feature set. By offering a PSTN-based system with an optional IP telephony gateway, AltiGen gives organizations the ability to simultaneously support PSTN and IP telephony and migrate traffic from one network to the other as appropriate.

5.2 Artisoft TeleVantage

Artisoft's Windows NT phone system, TeleVantage, is an excellent example of the sense of innovation that exists in the industry. With TeleVantage, Artisoft has blended their expertise in computer telephony and their background in user-friendly design to craft a Windows NT phone system that has become a market leader.

5.2.1 Target Markets

Just like AltiGen AltiServ, the target markets for TeleVantage are SMEs and departmental call centers. Currently, a single TeleVantage server can support up to 96 trunk lines and 256 extensions. In additional, multiple TeleVantage servers can be linked together over a LAN, WAN, or the Internet for greater scalability.

5.2.2 Client Application Features

With most PC-based phone systems, the majority of the discussion centers on the features and capabilities of the server software. Although TeleVantage provides a robust, feature-rich server, perhaps the most impressive feature of TeleVantage is the comprehensive and intuitive client application that is used for real-time graphical call control, managing voice mail messages, call screening, contact management, configuration, and more. The easy-to-use TeleVantage client software truly fulfills the promise of computer telephony through it's ability to give users the power to control, customize, and add intelligence to the way they communicate by phone.

The Artisoft TeleVantage Windows client software is an optional component, but it is strongly recommended for users who spend a large part of their day on the telephone. The client software provides a wide range of features including:

- A graphical interface to visually manage (place, transfer, conference, put on hold) multiple calls at the same time.
- A contact management database that allows users to place calls by clicking on a

contact name. This database is also used to provide screen pop notification of incoming calls.

- An interface for viewing, listening to, and saving voice mail messages.
- The ability to see the status of other users on the system.
- A call log that displays a complete call history and allows notes to be attached to a call.
- The ability to define personal call handling procedures and routing lists based on the identity of an incoming caller.

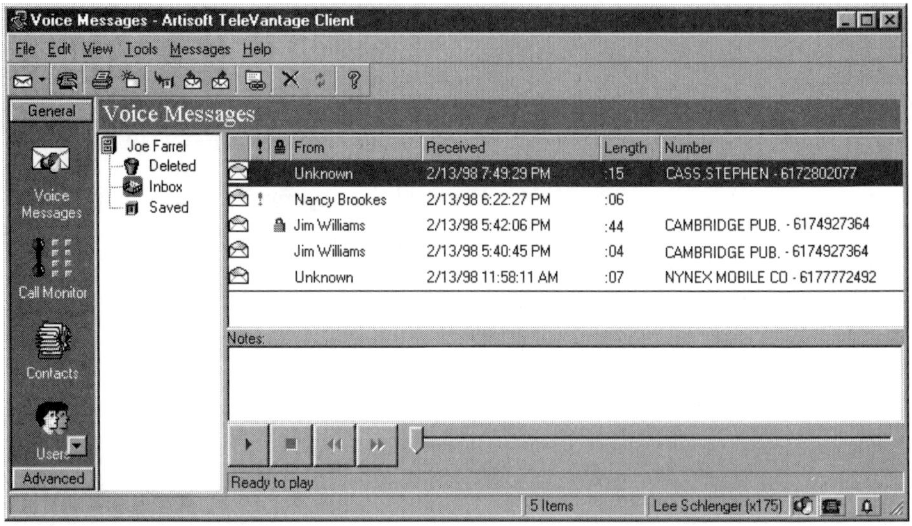

Figure 5-4: Artisoft TeleVantage Windows client software for managing messages, real-time call control, contact management, and more. Copyright (c) Artisoft Corporation.

By far, the most impressive features of the client software are the ability to customize the way calls are handled based on the caller's identity and the ability to easily handle multiple calls in real-time.

In addition to attempting to match the caller's phone number with one in the contact database, TeleVantage has two other means for establishing a caller's identity. TeleVantage can prompt the caller to speak their name, giving the user an opportunity to decide how to handle the call before answering it. Even if the user has directed the caller to voice mail, the voice mail message can be screened and the call connected if the user wishes. This entire process is controlled with TeleVantage's intuitive user interface.

Additionally, TeleVantage allows users to provide specific callers with subextensions that trigger a unique call handling procedure. For example, the owner of extension 101 has the set of subextensions from 1010 to 10199 at her disposal to assign to individual contacts or groups of contacts as she wishes. Each one of these subextensions can have its own prompts and routing rules. This is an excellent way to identify callers who should immediately be given a specific priority.

TeleVantage also includes a Web browser-based interface (see Figure 5-5) that allows users to manage their account configuration and an listen to their voice mail messages over the Internet. One of the authors has been using Artisoft TeleVantage as a remote worker for almost a year now. Between TeleVantage's sophisticated call forwarding features and browser-based interface, he has almost as much control over TeleVantage as any on-site employee—and much more than a remote worker of any other phone system would have.

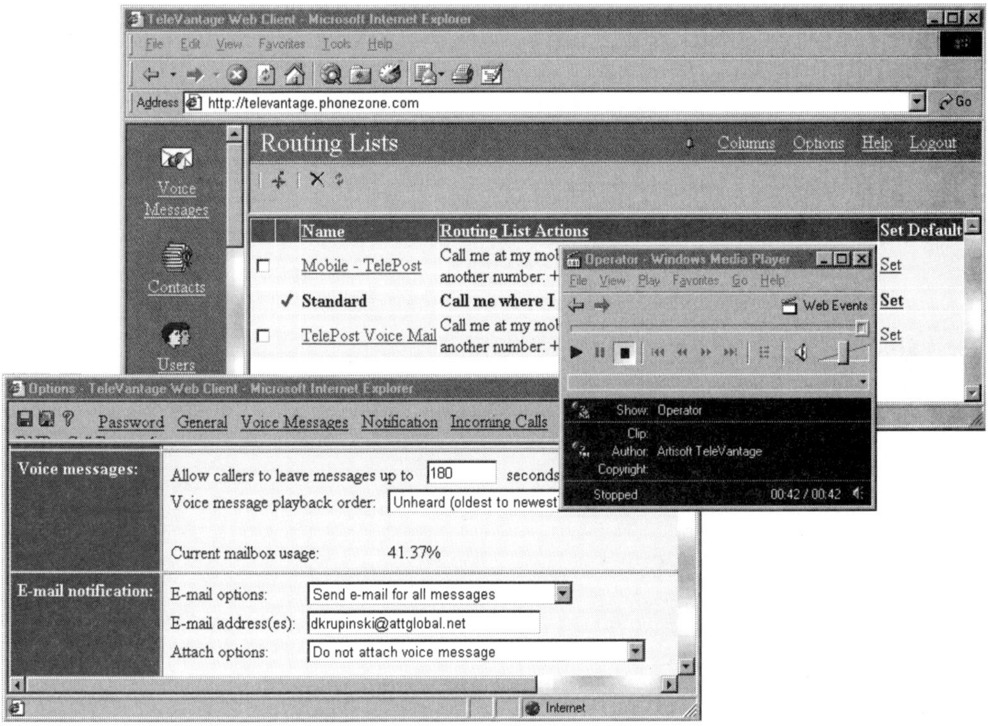

Figure 5-5: The TeleVantage Web browser-based interface for listening to voice messages and administering account options. Copyright (c) Artisoft Corporation.

5.2.3 Server Features

The TeleVantage server is designed to run on Microsoft Windows NT and use Dialogic analog and digital telephony expansion cards. TeleVantage server software controls the Dialogic hardware and provides a wide range of phone system services and features, including:

- Call switching and routing.
- Multi-level auto attendants.
- Voice mail, including the use of caller identification information to automatically return a call to the person who left the message.
- Sophisticated ACD support that includes three different routing schemes: round robin, first available, and broadcast. Additionally, TeleVantage allows ACD workgroup participants to be virtually anywhere. TeleVantage will treat an off-site phone number the same way it treats an internal extension. Accordingly, telecommuters and mobile workers can participate in ACD workgroups. Artisoft has also implemented overflow groups—secondary ACD groups that will only take calls if all the members in the first group are busy.
- A graphical administrative interface for configuring users and trunk lines, recording prompts, etc.
- International support which includes built-in voice prompts in Spanish, US and UK English (other languages can be added), flexible support for worldwide dialing rules, and support for a wide range of digital and analog telephony hardware certified throughout the world.
- E-mail and pager integration to notify users of new voice mail messages and to forward a voice messages as e-mail attachments.
- Call logging and detailed reporting for call centers.
- Integration with least cost routing services.
- H.323-compliant IP telephony gateway support.

Aside from the well-designed user interface, one of the most notable aspects of the TeleVantage server software is it's tight integration with the Microsoft SQL Server database. All administrative and call tracking information is stored in a SQL Server database that is installed with TeleVantage and licensed exclusively for use with TeleVantage.

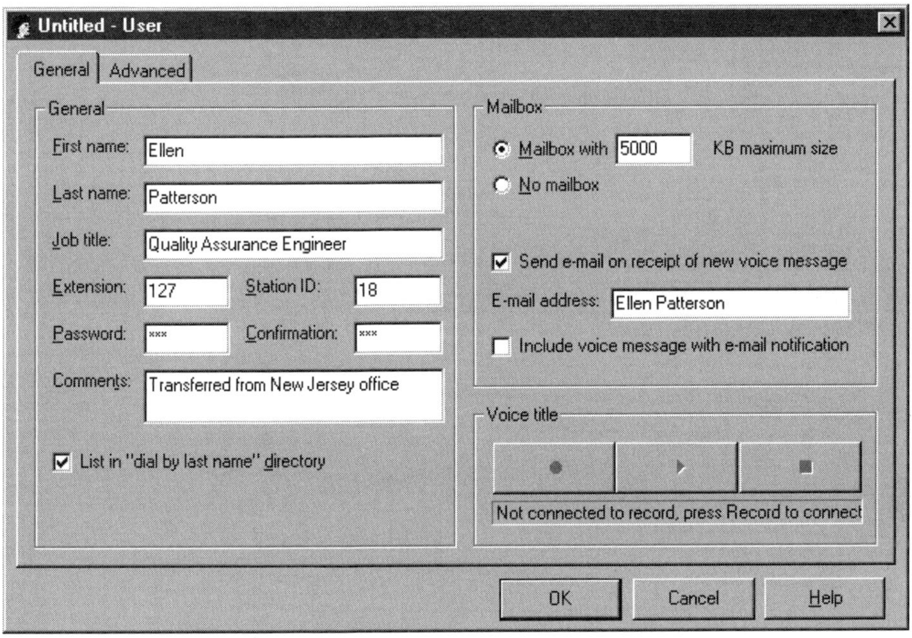

Figure 5-6: The Artisoft TeleVantage graphical interface for configuring users. Copyright (c) Artisoft Corporation.

The Dialogic Business Communications Platform

When TeleVantage was launched in 1998, it was the first phone system to take advantage of Dialogic's Business Communications Platform (BCP). The BCP is a combination of Dialogic computer telephony products designed specifically for use in PC-based phone systems. The BCP consists of an analog network interface board (with voice processing resources) for trunk lines, an analog station interface board for the extensions, and an external BCP connection panel.

The network interface board is a full-length ISA board available in 4 port and 8 port configurations, known respectively as the D80SC-4LS and the D160SC-8LS. What makes these boards unique is that they are *overloaded* with voice processing resources. Whereas a standard Dialogic analog network interface board would have the same number of voice processing resources as ports, the boards used in the BCP have twice as many voice processing resources.

The voice processing resources are used for a number of tasks, including analyzing the progress of outbound calls and generating tones (dial, busy, etc.) on the extensions. Accordingly, the number of voice processing resources is related to the number of extensions and not the number of trunk lines. In a system where the number of extensions is greater than the number of trunk lines, extra voice processing resources are required.

The station interface is provided by another full-length ISA board from Dialogic's MSI series. This board is connected to the network interface board via SCBus. In the future, we may see Dialogic integrate the network interface and the station interface onto a single board. Today, however, these two components are separate.

Perhaps the most unique aspect of the Dialogic BCP is the external connection panel. The connection panel is about the size of a laptop computer and is connected with cables to both the network interface board and the station interface board. Across the top of the panel is a grid of RJ-11 jacks for up to 8 trunk lines and 24 station lines. A power supply connects to the panel and provides the necessary voltage to "ring" the station ports.

Additionally, the BCP panel has a serial port connection to the PC. The connection panel expects to receive a "heartbeat" signal across this connection to verify that the server software is still operating. In the absence of such a signal, the BCP automatically redirects the incoming trunk lines to extensions. This is a unique feature of the Dialogic BCP specifically designed to address the reliability concerns of PC based telephony servers.

5.2.4 Summary

TeleVantage is an ideal solution for small businesses looking to add intelligence to the way they communication and enhance productivity. By providing a feature-rich client application, Artisoft has demonstrated that the value of "next generation" phone systems goes beyond the Windows NT server. TeleVantage provides users with a new way of looking at their phone system and offers users more flexibility to manage their communications than ever before.

INSIDER PERSPECTIVE

The Software PBX: Voice-Data Convergence for Today and Investment Protection for the Future

By Artisoft Corporation

Introduction

The shift from proprietary systems to open systems that occurred in the computer industry is now taking place in telecommunications, and businesses now have more choice in their selection of phone systems. Customers are buying communication solutions with complete application suites, lowering the cost of ownership of their phone systems, and deploying systems that will increase in value over time.

History Repeats Itself

The open systems market, which gives customers the freedom to mix and match technologies, provides purchasing power for companies that could not have been imagined during the early years of the computer industry. A customer had to purchase a complete system—computer, peripherals and software—from a single vendor. Once the customer made his purchase, he was locked in to a proprietary system that provided little flexibility. Technology choices were limited to whatever the single vendor could provide. And, because the customer was locked in to one vendor, the vendor could charge high prices for add-ons, expansions, upgrades and support. The cost of ownership for a business system was high because competition for add-ons, expansions, upgrades and support was non-existent.

Today, a business takes for granted that it can purchase a variety of computer software and hardware products, manufactured by a variety of vendors, through a variety of sources, and that these products will all work together as an integrated solution. For example, a business may purchase an e-mail system developed by Vendor A, a sales automation software application developed by Vendor B, a PC server manufactured by Vendor C, an extra disk drive manufactured by Vendor D, and some additional memory manufactured by Vendor E. Over time that business may upgrade one of the components, and will not be required to replace any of the other elements. For example, a business today may upgrade a server or replace a peripheral without replacing the related software application.

Looking back, it's clear that the shift towards open systems has brought dramatic benefits to the computer industry. It has increased competition in almost every area of technology, which has allowed new technologies to emerge at a tremendous pace and caused prices to plummet. It also changed the face of the industry, as new companies providing open systems products emerged as leaders over more established vendors.

Shift to Open Systems

Driven by accelerating requirements for business applications and increased functionality, the phone system industry is now undergoing a similar shift towards open systems. And the benefits for businesses are even more dramatic.

Historically, a business had no choice but to purchase a proprietary phone system, where all the components—switch (PBX), handsets, peripherals, and applications—came from a single vendor. Often, a business had to abandon its communication systems as it grew, or was dependent on a single supplier to meet its technology needs. As a consequence, growth was often interrupted or slowed by the business' inability to grow the phone system.

Over time, business phone systems have become even more mission critical because of their integration with other systems. When a business interacts with its customers and suppliers, its information technology (IT) systems shapes those interactions. The company's Web server, e-mail, customer relationship management software (CRM), and accounting software are often integrated with the PBX, the automated attendant, the interactive voice response (IVR) system, and voice mail. These systems come together during customer interactions: A customer calls, finds the right person to talk to, product information is provided, an order is entered, a bill paid, a service contracted for.

Today, PBXs are part of the fabric of the way a company operates, and this requires flexibility in hardware and software interfaces. Internet connections change and get faster. Web servers have to be upgraded to accommodate exploding demand. Storage subsystems increase in sophistication, and can take Web-oriented applications up into capacity ranges that had been exclusive to mainframes. Desktop PCs have a limited lifespan due to constant technology improvements. In this changing environment, a business cannot afford to be locked into an inflexible PBX environment.

The Software PBX

Fortunately, the telecommunications industry is moving away from proprietary systems and is embracing open systems. And, as happened in the computer industry, the introduction of open systems is providing customers with significantly better products and greatly reduced costs.

The key technology that is enabling the phone system industry to move to open systems is the software PBX. A software PBX requires no proprietary hardware; instead, it uses industry standard interfaces and runs on commercial, off-the shelf hardware. With a software PBX, the phone system has become another set of software applications (call manager, voicemail, auto attendant, ACD, etc.) and another set of peripherals (handsets, headsets, IP devices, etc.) running off a standard computer server. The software PBX is the result of an evolution that has occurred over the last ten years, driven by an important trend—the convergence of voice and data technologies.

Convergence of Phone Systems and Computer Systems

Historically, businesses that required a phone system chose from several proprietary PBXs. These systems combined purpose-built hardware with closed and proprietary software that had little or no outside interface other than network communications protocols for incoming and outgoing PSTN calls. Today, vendors continue to sell these systems because they are lucrative and lock in a customer to a single source of hardware for the long term.

Certain telecommunications manufacturers began to recognize this underlying need for openness, and began to develop and manufacture products that attempted to uncouple the software from the hardware, and take advantage of computer system standards. The convergence of phone systems and computer systems has evolved over four phases:

- Computer Telephony Integration (CTI)
- Proprietary PC PBX
- Proprietary IP PBX
- Software PBX

Computer Telephony Integration (CTI)

About 20 years ago, the first standards-based connections to PBX systems were introduced. Computer telephony integration (CTI) brought standard control interfaces, such as the ECMA PBX-to-host link, to proprietary PBXs, mostly for the purpose of interfacing voice processing systems such as voicemail and interactive voice response (IVR). At this phase of evolution, PC-based voice processing systems worked alongside legacy switching hardware.

Serial connections and simple standard protocols enabled call-center software and voice mail systems to communicate with PBXs.

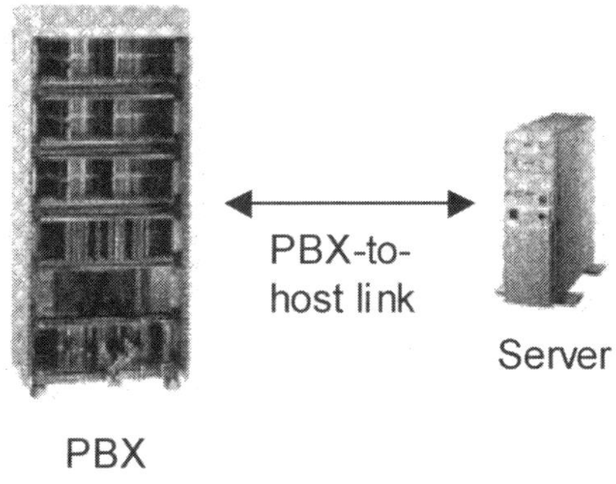

Even though standards were implemented in numerous variations, the initial CTI systems had some advantages: PBXs no longer sat isolated, and interfaces were no longer under the control of a single vendor. Still, the cost of adding CTI interfaces to legacy PBXs was, and remains, high. Customers still had the expense and limitations of the proprietary PBXs, and acquired additional integration and maintenance issues associated with the new server. Further, the relatively narrow and inflexible

CTI interfaces—typically low-speed serial links—were often insufficient for new tasks beyond the limited feature set, such as basic transfers of calls.

PC PBX

About four years ago, the first PC-PBXs started to appear. In these systems PCs, which had previously just handled voice processing, handled some call switching as well. This had some advantages:

- PC operating systems are always increasing in power and capability. PC PBXs were able to inherit some of these capabilities.
- PC PBXs could run application software in the same box with the switching hardware. This eliminated some of the restrictions that CTI interfaces to proprietary PBXs have.
- PC PBXs sometimes came with an applications suite that provided messaging and other features only found in expensive legacy PBX/voicemail combinations.

However, as development continued, it became apparent that the PC PBXs on the market continued to tie proprietary telephony hardware in a PC to the software. PC technology continues to evolve very rapidly offering tremendous advantages over its predecessors. However, the inflexible, proprietary telephony hardware of a PC PBX prohibits you from deriving the benefit of newer technologies since the proprietary telephony hardware of the PC PBX cannot keep pace with the speeds and feeds of the rapidly evolving PC. This problem is very clearly illustrated by the evolution of the PC expansion bus interface.

The adoption of the PCI bus by the PC industry clearly demonstrates the disadvantage of a proprietary PC PBX. Businesses that purchased PC PBXs began to see compatibility issues with ISA cards and PCI slots. Although the PCI bus was a step forward in computer evolution, the PC PBX did not enable easy migration to newer, faster platforms with newer, faster bus interfaces. The simple task of increasing the number of stations on a PC PBX system became very difficult and expensive—requiring businesses to acquire old PC technology or completely replace their system. The PC PBX did not eliminate dependency on a single or a small number of hardware vendors.

IP-PBX

As Voice over IP technologies improved, several vendors began to take the approach of shifting all communications to a business IP network. IP PBX vendors introduced new network technologies, which would control a network of proprietary IP telephones. Because IP standards were still evolving, each vendor interpreted the standards differently or ignored the evolving standards completely thus creating telephones and routers that were incompatible with competing vendors. Thus the telephones, routers and hubs introduced with these systems remained proprietary, and were often extremely expensive. Further, very basic application sets were provided, with little flexibility for other application developers to develop custom applications that leveraged the advantages of converged voice and data network. Instead the IP PBX merely replicated the limited feature set of traditional proprietary PBXs within a data-centric environment.

Software PBX

Several years ago, after numerous experiments with combining computers, networks, and PBXs, including CTI, PC PBXs, and IP PBXs, the industry finally saw the introduction of a true software PBX. The software PBX takes advantage of a comprehensive set of standards that enable software to control switching hardware and voice processing hardware without becoming entangled in implementation details of the hardware. This enables underlying hardware, signal processing, and communication links to be changed without forcing the user to discard the part of the system that defines the way voice communication works at his enterprise. A software PBX does not require proprietary hardware, since the software is compatible with industry-standard telephony hardware and off the shelf servers. It provides plug-and-play compatibility between software, hardware and all peripherals, allowing great flexibility in implementation.

Advantages of Software PBXs

A software PBX will not become obsolete. When a business implements a software PBX, the business has invested in a product that will allow the company to upgrade the hardware in the future without losing the investment in the software functionality and productivity improvements. Conversely, businesses will be able to upgrade their software to take advantage of new features that more tightly integrate their data and voice infrastructures without having to replace their hardware.

Standard hardware interfaces enable systems to start small and scale up to very large multi-location networks. Standard driver interfaces enable multi-vendor hardware solutions. Flexibility in hardware options provides investment protection into the future, as new less expensive technology is proven in the marketplace.

A software PBX provides infrastructure flexibility. Each business has a unique infrastructure, and a variety of business requirements. Using standard network interfaces that enable connectivity to the conventional telephone network and packet telephony systems and handsets, a software PBX can be implemented in a traditional PBX environment, integrated with the IP network, or in a distributed PBX/branch office environment. A business implementing a software PBX has the flexibility to evolve at their own pace to implement VoIP, or adopt a hybrid approach.

A software PBX lowers the cost of ownership of communication systems. Once a business has implemented a software PBX, future capacity upgrades will be straightforward and inexpensive, whether software upgrade or hardware upgrades. Competition between developers and manufacturers will result in price reductions, and simplified upgrade procedures lead to lower service fees. Because a business is no longer dependent on a single PBX vendor, the IT buyer does not have to manage supplier risk, and achieves greater bargaining power among multiple suppliers. Leveraging competition among suppliers results in more options and choice, as well as lower cost of ownership over time.

A software PBX allows a company to embrace new technologies quickly, as they become proven in the marketplace. With a software PBX, upgrades to capacity, availability, connections, and storage are all as transparent to the users—to the people whose productivity the system is meant to enhance—as upgrades to the underlying platforms for other enterprise applications. The phone system can become more tightly integrated with other systems—order processing and IVR systems, Web-based customer care systems, using capabilities such as VoIP and remote call routing.

A software PBX provides more choice in features and applications, and a more customizable system. The software PBX can integrate with voice and data applications using standard system API interfaces and Microsoft's COM distributed component technology in the Windows 2000 environment. Applications that are customized to the requirements of the business can be developed using standard APIs. For example, unique CRM applications with screen pops, paging applications, and call center applications are easy to implement and modify as the requirements of the business evolve.

Artisoft's TeleVantage Software PBX

Artisoft's TeleVantage is the leading software PBX available today. TeleVantage is designed to take advantage of open software standards, and to provide infrastructure and applications flexibility for business customers.

By adhering to industry standards, TeleVantage is independent of hardware and provides maximum flexibility in choosing hardware suppliers. TeleVantage interfaces to applications on other platforms using Windows standard telephony APIs for telephony-specific applications and COM for network-distributed integration with enterprise software.

TeleVantage Infrastructure Flexibility

Because Artisoft's TeleVantage has been developed using industry standard interfaces, it affords a business great flexibility to connect telephones, phone lines, data

Layer	Description	Benefits
TAPI API	Windows 2000 API layer for	Applications that adhere to Windows 2000 API telephony applications standards will run on TeleVantage. Also, because TeleVantage itself runs on Windows 2000, it can implement more TAPI features and provide better performance than systems that are not PC-integrated. This interface is for applications like voice mail and IVR.
MAPI	The Windows messaging API	Using MAPI, TeleVantage is compatible with the widest possible range of e-mail servers and e-mail client software.
ODBC	A Windows database API	Using ODBC, TeleVantage integrates with SQL Server for logging and other internal databases.
COM	A Component Object Model (COM) interface to the internals of TeleVantage	The TeleVantage COM interface provides a very powerful, yet high-level and object-oriented interface to TeleVantage internals. This interface is appropriate for interfacing TeleVantage to other enterprise soft ware systems, like accounting software or Web applications. An example of a use of this interface is Artisoft's own Web-based interface to system features. This example shows that every feature of the product can be accessed by Web applications.
S.100 Middleware	Developed by the ECTF, S.100 is an industry standard API for controlling computer telephony (CT) resources including voice processing hardware, switching hardware and speech technologies. S.100 encompasses the underlying middleware that implements the API.	Artisoft is committed to the S.100 standard and has announced a joint development project with Intel to implement it. As TeleVantage supports S.100, businesses using TeleVantage will simply upgrade their software to take advantage of this new standard. Every manufacturer that supports S.100 is a potential supplier of hardware for CT Servers that adhere to this standard.
S.300	A service provider interface for hardware drivers–think of it as the underside of S.100	S.300 enables multiple manufacturers to write driver software for their hardware that talks to any S.100 middleware.
H.100	A hardware interface for voice processing cards	H.100 enables multiple manufacturers' voice processing cards to plug in to the same chassis and connect to each other. This enables future versions of TeleVantage to offer a very wide range of telephone and network connections, and a high degree of scalability as hardware increases in capacity.
H.323	A protocol standard for packet telephony on IP networks	H.323 enables TeleVantage to connect to IP telephones, and for PCs running NetMeeting and other H.323 software to act like telephones. It also enables TeleVantage servers to connect to public packet telephony gateways, and for multiple TeleVantage servers to be linked in large virtual systems.

Artisoft's TeleVantage provides standards compliance at each application layer.

networks, network service providers, and the distributed parts of an enterprise at distant locations. The following figure shows how TeleVantage provides a very broad set of connectivity options:

Standards make connectivity choices possible. A wide range of connectivity choices are illustrated below:

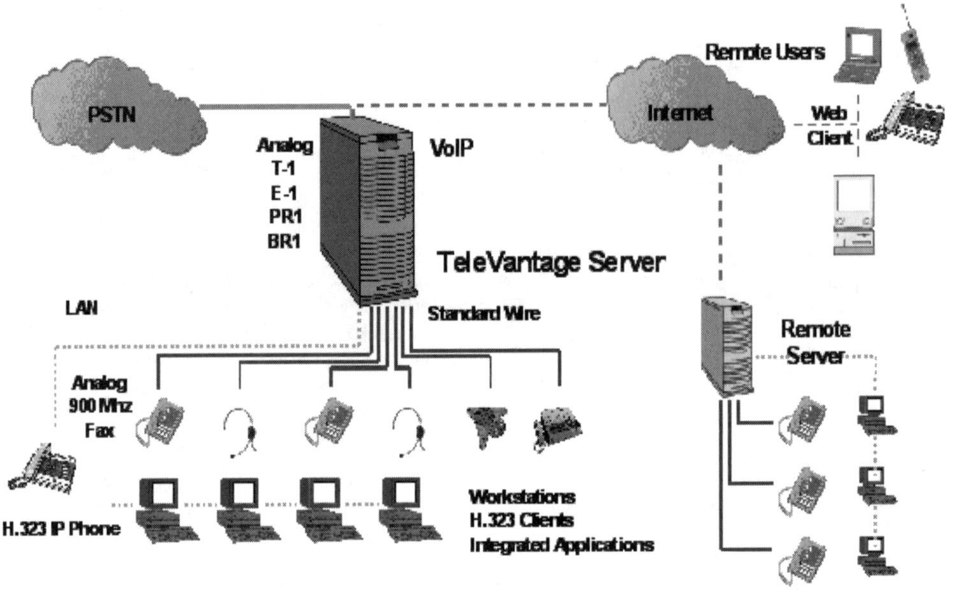

- Digital and analog trunk connections to the public switched telephone network (PSTN), including analog, T-1, E-1, Primary Rate ISDN (PRI-ISDN), and Basic Rate ISDN (BRI-ISDN)
- Voice-over-IP (VoIP).
- Remote connection to off-premises extensions (OPXs).
- Remote connections to Web clients.
- Connections to remote servers to create large virtual systems.
- Local connections to a variety of conventional telephones including conference speakerphones, wireless phones, and call-center-oriented headset-equipped phones.
- Local connections to IP phones over existing data network.
- Local connections to PCs running H.323-based conferencing software like NetMeeting.

Future connectivity options will provide additional choices for IT managers, and never require a decision to replace a whole system. ISDN phones, digital wireless phones, wireless PDAs, and other devices will be able to connect locally, and convergence technologies like Voice-over-DSL (VoDSL) will enable a single Windows 2000 server to provide data routing and connect to voice trunking on a single connection from a telecommunications service provider.

TeleVantage Applications Flexibility

TeleVantage provides a robust turnkey set of business applications, including PBX functionality, automated attendants, ACD capabilities for call centers, and voice mail integrated at the desktop. The following examples are elements of the Artisoft TeleVantage applications suite:

A call-control application provides a graphical user interface to telephone operations, putting powerful operations like call transfer and conference calling at a PC user's fingertips. Multiple calls can be easily manipulated with an on-screen interface.

A visual interface to voice mail provides fast PC-based access to voice messages. Voice messages can be treated like e-mail; they can be organized, prioritized, and referred to over time for tremendous productivity improvements.

Call and message screening allows an employee to manage every day voice transactions in priority order. Call handling can be personalized so that employees can be reached at office, home, and mobile phones, according to the user's preference. On the basis of caller-ID information, different callers can be handled differently.

Automatic call distribution (ACD) is the basis of call-center functionality for e-commerce businesses. TeleVantage provides a powerful ACD capability along with a visual interface that makes setting up ACD functions easy.

Logging and reporting software make it easy to see that a TeleVantage system is functioning correctly. Logging and reporting call center activity is particularly useful because information on hold times and call volumes have a direct impact on e-commerce effectiveness.

Remote users can take advantage of productivity improvements when not at their LAN-connected desktop. For example, using the TeleVantage Web interface, a remote user can take advantage of a cell phone and by contacting a company system, build a conference call.

Extensible Programming Interfaces Because TeleVantage applications have been built using Microsoft COM interfaces, integrating TeleVantage with e-commerce Web sites is a straightforward extension of this functionality.

Summary

A software PBX has a powerful set of features that make excellent use of the network of PCs that is fundamental to a company's operations. It provides more value than a proprietary PBX because it spans conventional and packet telephony, and incorporates features like ACD and extensive reporting that would be costly options or separate products in other systems.

Web sites, call centers, and back-office software systems become ever more integrated as companies pursue greater efficiency, global reach, and new capabilities through e-commerce. Voice communications are part of the e-commerce environment, and they should be a well-integrated part. A software PBX is managed as a component of a company's Internet-oriented software infrastructure, and integrates seamlessly with the company's voice and data networks.

Acquiring a software PBX is a prudent decision that unchains voice communications from hardware platforms. Underlying hardware, network connections, and network connectivity can be changed or upgraded in cost-effective ways that are transparent to business operations. Software-PBXs provide a better value today and a technology-independent investment for the future.

This Insider Perspective was contributed by Artisoft, Inc. Many thanks to Greg Gies and the folks are Artisoft for their assistance. Within these pages Artisoft introduced a new category of next generation phone systems—the Software PBX. This is yet another testimony to the versatility, flexibility, and power of the programmable PC platform.

5.3 COM2001.com InternetPBX

The COM2001.com InternetPBX is one of the most feature-rich PC-PBXs on the market today. In this Windows NT-based phone system, you'll find all of the features in most other PC-PBXs, plus a voice recognition enabled auto-attendant, a conferencing bridge that supports up to 64 parties, and unified messaging support that includes a text-to-speech e-mail reader.

5.3.1 Target Markets

With this solution, COM2001.com is targeting fast growing small businesses—especially those with sophisticated communication needs—and departmental call centers. Today, its Internet PBX features sizes range from 12 ports (supporting 4 trunks and 8 extensions) to 144 ports (supporting 48 trunks and 96 extensions).

Figure 5-7: The COM2001.com InternetPBX toolbar provides users with a graphic interface for placing calls, answering calls, setting up conference calls, and more. Copyright (c) Comm2001.com.

5.3.2 Features

According to COM2001.com, the features its InternetPBX include:

- Integrated voice over an IP gateway allows users to mix standard and IP phone service, depending on call destination and desired quality. IP telephony is especially suited for low-cost calls to other offices, fellow employees, and preferred vendors.
- Unified messaging gives users one easy-to-access location for all messages. The InternetPBX places voicemail, e-mail, and fax messages in a Microsoft Exchange database. Users can easily view, save, delete, forward, and store these messages on their computer desktop. Unified Messaging reduces the time and effort of checking messages and the risk that an important message from a customer will be lost.

- High-quality digital conference bridge enables users to set up conference calls for up to 64 participants with a few clicks of the mouse. Participants enjoy excellent sound quality, and the InternetPBX Toolbar lists the names and contact information for every participant.
- Auto attendant with advanced speech recognition routes calls directly to the person or department sought. It provides a dial-by-name and dial-by-extension listing directory to give callers rapid access.
- Interactive voice response (IVR) system includes an integrated speech recognition and text-to-speech application generator. Specifications and results can be programmed to enable callers to obtain desired information automatically.
- Personal assistant server — Alexis — greets callers, asks who they are calling, and then connects them if authorized. Alexis uses "Find-me Follow-me" services to locate users who are not in the office. Alexis also responds to voice requests from users to look up addresses and phone numbers and to place calls. Alexis can even use a text-to-speech processor engine to read e-mail to a user over the phone.
- Desktop toolbar enables users to place, accept, transfer, send calls to voicemail, or forward a call or message with the click of a mouse. Thus users can perform all their communications tasks more quickly.
- Single number access means that users have to provide only one number for their office, home, mobile, and fax phones. They can instruct the InternetPBX system to forward calls to any other number that will reach them, or even to try a list of possible numbers.
- Caller ID screen pops let users see who is calling, sometimes before the phone even begins to ring. Users can choose which calls to take and which to send to voicemail or forward to a more appropriate person.
- Virtual office over the Internet gives users full access to their phone and messaging services in real-time through a Windows NT-authenticated and secured Web-based server that can be accessed from any location. Screen pops and call control features are programmed in HyperText Markup Language (HTML) for real-time remote access.
- A virtual private network (VPN) for each user allows every user to access phone services from virtually any location, using a variety of InternetPBX network and connection strategies, including LANs or WANs, the Internet, and standard and wireless phone networks.

In addition to the array of features packed into InternetPBX, the other unique aspect of this solution is the tight integration of InternetPBX with a wide range of Microsoft applications. For example, InternetPBX uses the Microsoft Management Console (MMC) for the administration and configuration interface. This is the same

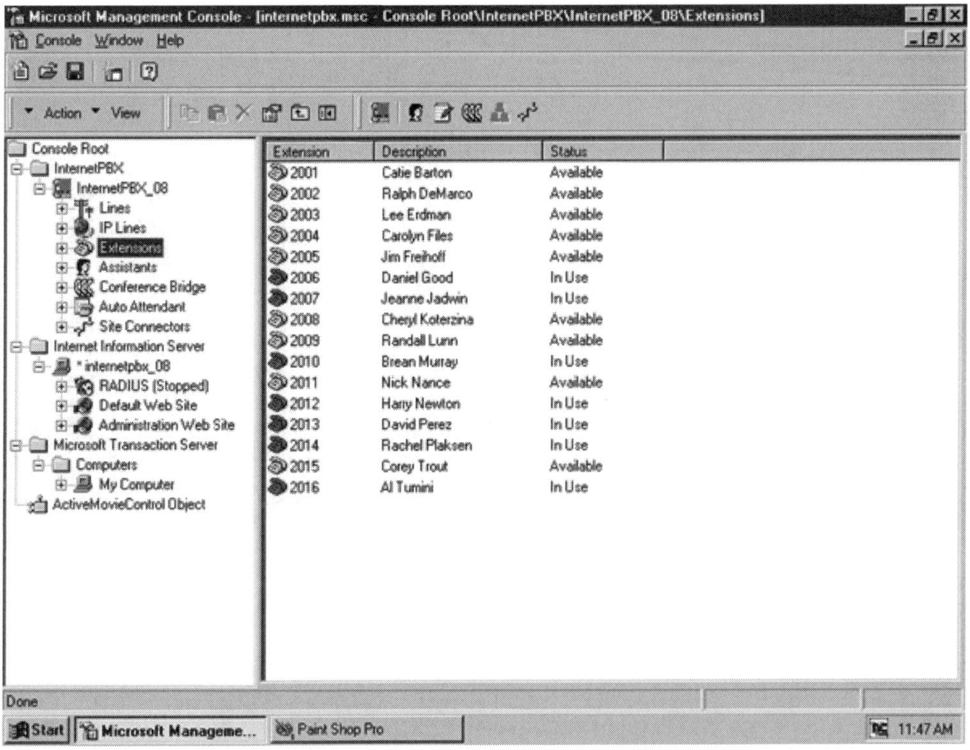

Figure 5-8: Using the Microsoft Management Console to configure InternetPBX and other Windows applications. Copyright (c) Comm2001.com.

administration interface used by many Microsoft applications, including Internet Information Server (IIS), Microsoft Transaction Server, and others. InternetPBX modules utilize the snap-in architecture of the MMC to provide a common administrative interface for InternetPBX and other Windows NT applications.

InternetPBX is also integrated with the Microsoft Exchange Server for message storage. InternetPBX uses the Exchange Server database to store all voice, fax, and e-mail messages. Accordingly, InternetPBX users can retrieve all of their messages, regardless of type, from their Microsoft Outlook desktop client. Plus, the InternetPBX personal assistant, Alexis, provides access to voice and text messages to users over the telephone. Using text-to-speech technology, Alexis can convert e-mail messages to speech and speak them over the phone. The COM2001.com InternetPBX provides one of the most complete unified messaging solutions of any PC-PBX on the market today.

Microsoft Outlook users are also able to configure InternetPBX to integrate with their Outlook contact database. With this feature, InternetPBX will attempt to identify an incoming call by matching the incoming CallerID with the information in the Outlook contact database. In addition, InternetPBX users can easily place a call to any of the contacts in their Outlook database through the InternetPBX toolbar.

5.3.3 Summary

The COM2001.com InternetPBX packs an impressive collection of features in a PC-PBX platform. The solution exploits the programmability and interoperability of the PC platform to provide one of the most comprehensive solutions on the market for businesses that want to communicate more efficiently and enhance their productivity.

INSIDER PERSPECTIVE

Internet PBX: A Powerful New Phone and Messaging System for Growing Companies and Businesses of All Sizes

By David Perez—President and Co-Founder, COM2001.com

Software-based InternetPBX replaces traditional business phone systems with a product that consolidates analog and digital features-traditional telephone service, e-mail, voice-mail, and faxing integrated with low-cost Internet Protocol (IP) telephony. Users gain advanced phone services that help them deal with the challenges of business communications today.

Benefits of InternetPBX include:

- faster and better communication with customers and suppliers;
- dramatic increases in organizational productivity with improved services;
- low cost long distance with IP telephony;
- a system that's easier to install, manage and expand.

The Problem with Business Communications Today

Business communications today face three problems. The first is sheer volume. Communication seems to grow constantly in most companies. Knowledge workers are overwhelmed. It is more difficult to keep up with phone calls and e-mail. Critical customer communications are often buried or lost in an avalanche of information.

The second problem is that messaging now comes in many different forms via many separate systems. Workers use landlines and wireless phones to receive phone calls and check voicemail. They must go to a fax machine for faxes (too often on a different floor or even in a different building). They receive e-mail on their computers, and often must go online to check for mail.

Messages in different formats are often redundant. Business users must access several systems to find customers' messages, and they often must respond multiple times to a single concern. For example, when a customer could not reach a support engineer on the phone, the alternative was to page that person. When no one answered the page, the customer sent an e-mail or a fax. So the engineer received the same message through several different systems.

The third problem is change. Communication technologies and systems for business are continuously changing. Even the businesses are constantly changing-upsizing, downsizing, and reconfiguring. The cost of changing a traditional phone system multiple times can be larger than its initial price. Older analog systems or those built with proprietary technology generally have closed architectures that cannot be upgraded or cannot have new kinds of services added efficiently. COM2001.com designed

InternetPBX specifically to address the need for improved communications productivity. To increase efficiency, it focused on creating systems that work the way

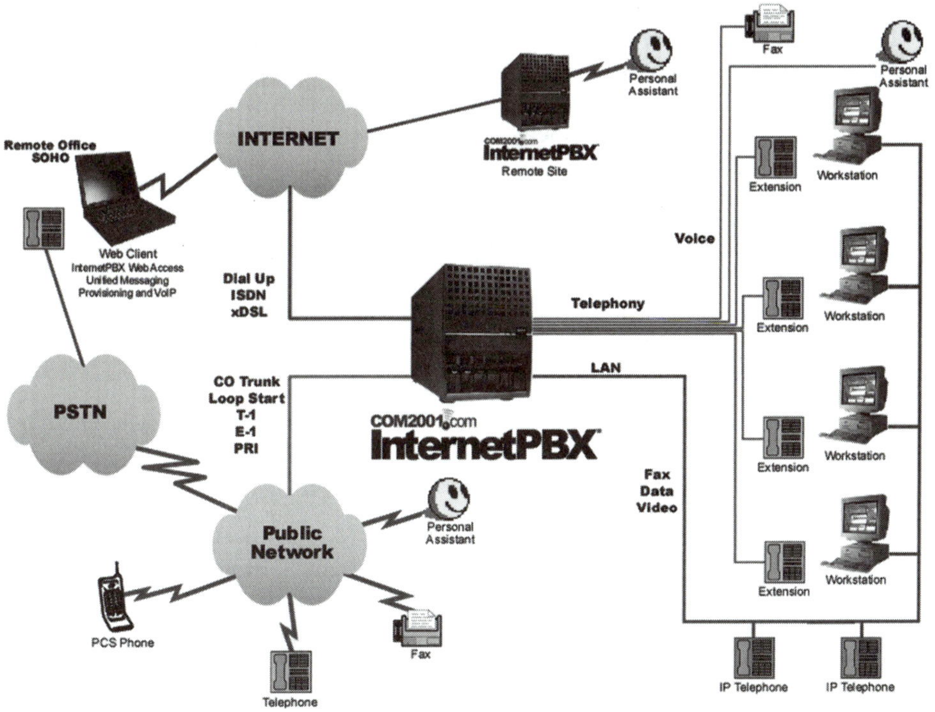

users work. To deliver more affordable solutions, it focused on rapidly converging computer telephony (CT) and IP telephony technologies.

What Is InternetPBX?

InternetPBX defines a new kind of business phone solution. It replaces traditional business phone systems with more versatile PC-based technology that unifies phone, messaging, and productivity services in one solution, based on the convergence of voice and data within digital networks.

It is built on open standards, using robust and ubiquitous PC technology platforms (Microsoft and Intel) and off-the-shelf components. Its IP telephony provides lower costs for business users. InternetPBX provides virtually every phone and messaging service small businesses need, as illustrated below.

Why Businesses Should Install InternetPBX

InternetPBX provides businesses with more than a simple increase in number of features and services. The most important reason is improvement in the speed and quality of communications in both directions.

- Easier to reach employees—Important callers like customers and suppliers find it easier to reach employees. The single-number, "Find-me Follow-me" service, and the Alexis Personal Assistant help to keep customers connected and to increase the number of successful connections with users, even when they are on the road. Users can return calls quickly and access e-mail while on the move.
- Improved call control—Because users can see who is calling, often before the phone even rings, they can prioritize their calls, answering the most important ones and sending the others to voicemail. Listing voicemails in the user's personal Microsoft Outlook database makes it easier to respond to voicemail. Voicemail can be played and calls returned by voice or e-mail, all with a single mouse click. An integrated Operator's Console lets users see and manage all incoming calls and view activity on all other user extensions.
- Additional ways to communicate—Users can easily and quickly decide whether to return a message by e-mail or voice. Both are equally easy when the party is listed in the user's Microsoft Outlook database.
- Increased productivity—This results from better communication with and control of calls from both external and internal customers. The advanced user interface and tools help users to miss fewer calls, receive all their messages-voicemail, e-mail, and faxes-from a single system, and return calls faster. They waste less time playing phone tag. A Personal Assistant can forward calls and seamlessly connect the caller to any destination or device (wireline or wireless phone).

- Reduction of costs for long distance and international calling—Many providers are using IP solutions to reduce the costs of providing services. InternetPBX was designed to give users direct access to IP telephony-especially for long distance and overseas calling between offices. The cost of IP telephony using an existing Internet connection is almost zero.
- Easy customization of services—The Microsoft-based architecture of InternetPBX ensures easy customization of services for dedicated segments. Since the software is built on open industry standards for both data and telephony net-works, it can be easily integrated with third-party applications. Both COM2001.com and third-party providers can easily support specialized functions such as distributed call centers and service bureaus.
- Lower capital costs—InternetPBX is substantially less expensive than traditional, proprietary, voice-only telephone systems in terms of capital costs to install, integrate, and change services. Installation takes a few days because InternetPBX uses existing network wiring and can operate with virtually any handset. Services or locations also can be easily added because of the flexibility of using almost any kind of networking; it is similar to adding one more service to a company LAN or WAN.
- Investment protection—The COM2001.com integration strategy helps to protect customer investments and ensure that they will always have up-to-date technology. InternetPBX is built from robust and mainstream technologies. These proven hardware and software technologies, essentially off-the-shelf components, are continually upgraded by vendors such as Dell, Intel, and Microsoft. The InternetPBX Support Plan gives customers the opportunity to stay current with best-in-class price/performance.

Tracking the Growth of Enabling Computer and IP Technologies

Since the invention of the personal computer in 1970, the record shows that the opportunity in every technology is maximized in the digital world where price/performance doubles at least every 18 months. (This is compared to the electromechanical world, where improvement averages perhaps 10 percent per year.)

Opportunity is increased for solutions that involve networks as a multiplier. And it is increased even further when there are open architectures and standards with multiple suppliers of enabling technologies. Products based on closed architecture and proprietary standards develop slower and therefore tend to deliver less value to customers. The successful products of today and tomorrow are digital, networked and open.

Older business phone systems suppliers come from a deep tradition of closed and proprietary technologies. Companies that evolved from the Bell System, and even those that competed with them, still sell systems that are closed and proprietary at their core. These systems typically require bridges and gateways to connect to today's open systems and standards. Bridges and gateways also add cost, limit flexibility, and slow innovation.

PC technology, based on open standards set by IBM in 1980, rapidly outgrew IBM's "big iron" of mainframes and mini-computers. Even today's super computers now use PC chips, although in massively parallel arrays. Today a single PC with multiple Intel processors can contain more computing power than the entire Bell System in 1950.

The Internet, while it depended on proprietary telephone networks to transport its open and standard protocols with digital data, grew faster and built significantly more value than the powerful telephone network companies.

Voice services were protected both by proprietary standards and by legislated tariffs. Increasingly, however, data made up more of the traffic on telephone networks designed for voice circuits. It was demonstrated in the 80s that voice could be transmitted more efficiently in digital form in the switched packet technology of the Internet. More small, switched packets of data-encoded voice can fit in a network "pipe" than the bigger, less efficient, switched analog circuits that traditionally carry voice.

Sending voice by Internet is called IP Telephony. The Internet is in many ways free or certainly low in cost. So is IP telephony, and its use is continually increasing. There are still problems of quality, because the original Internet Protocol was not designed for the requirements of two-way voice communication. But these are being solved rapidly. Today it is recognized that most communications—voice, video or data—will be in IP format at some point in their transmission.

When companies look for ways to lower their voice communication costs, IP telephony is always part of the solution. Telephone companies like AT&T are moving voice traffic to IP telephony between gateways, but move it back to traditional telephone protocols for delivery to homes and businesses. Price pressures will force them to feature IP more in the future.

E-mail is now used more frequently than voice messages or even "snail mail". Growth of e-Mail is estimated by the Business Communications company to be growing at least 42 percent per year through 2002 At least part of this growth is caused by the ease of sending copies of e-mails to others. A member of the Internet Engineering Task Force recently stated that fax services now take up increasing amounts of network transport space. Soon digital (streaming) video certainly will be an additional factor.

Different sources cite different figures, but it is clear that data traffic is passing voice traffic on networks in terms of sheer volume. In the near future, voice will be transported primarily in digital data format. Voice is moving quickly from traditional proprietary systems to systems with open architectures.

The 1999 Multimedia Telecommunications Market Review and Forecast, a co-publication of the Multimedia Telecommunications Association (MMTA) and the Telecommunications Industry Association (TIA) says that in 1992, 82 percent of all computer telephony system products were proprietary. By 1998 as many as 85 percent of products were built on open architecture. In addition, the "open telecommunications PBX market" is expected to experience 116% compound annual growth rate between 1998 and 2002. (F.A.C./Equities research organization.)

Open PBX Market Growth Projection

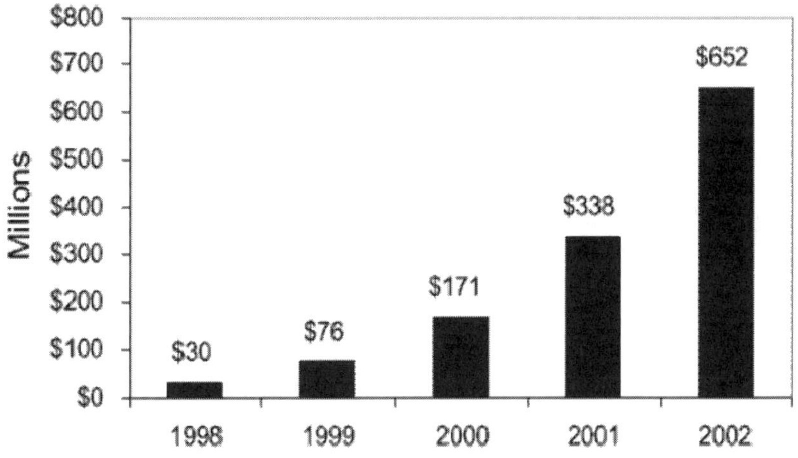

Source: FAC/Equities

Costs of computer telephony solutions, especially driven by increased use of IP telephony, continue to drop. The average price of CTI stations for call centers has dropped from as much as $3,600 a seat (for hardware and software) in 1993 to as little as $1,300 in 1998. But it is important to remember that computer telephony solutions have other ways of adding productivity and value.

Computer telephony solutions can bundle "unified messaging" so that essentially all forms of communications—phone, voice mail, e-mail, paging, and fax—can be delivered by the same system. This provides a huge opportunity for increasing productivity.

All of these technology opportunities—unified messaging, converged voice and data networks, IP telephony, and PC-based telephony—are coming together to create new categories of business phone solutions.

InternetPBX for Small, Growing Businesses

InternetPBX was designed for small businesses, businesses with mobile workers, and especially businesses that are experiencing rapid growth. It provides the flexibility and extensibility those companies need but cannot get from traditional business phone systems.

For example, sales departments can place the InternetPBX server between their legacy PBX and their customers, obtaining improved services and features without causing an enterprise-wide overhaul of telephony services.

Individual users save time with the unified messaging solution provided by InternetPBX. Recent productivity studies, for example, indicate that executives may save as much as 27 minutes per day by obtaining improved access to e-mail and the ability to work while mobile. Individual users benefit from InternetPBX whether they are on the road or in the office.

- InternetPBX on the road—While waiting in traffic or even on an airplane, users can review, act on, and forward e-mails, voicemails, and faxes by using the text-to-voice Personal Assistant over their cell phones. When traveling, users can connect to the Internet using a dial-up connection and a standard browser such as Microsoft Internet Explorer or Netscape Navigator. An incoming phone call to the home office instantly prompts the InternetPBX to display a Screen Pop on the user's laptop showing the caller's ID and asking where the call should be transferred. Calls can be accepted remotely, transferred to another person, or sent to voicemail.
- InternetPBX in the office—InternetPBX gives users a unified look at all voicemail, e-mail, and faxes in their Microsoft Outlook inboxes. The Busy Lamp Field of the system allows all users to see the status of all other users (in or out of the office,

available or unavailable). Users can save time by waiting until other users are available before they try to communicate with them. They need not waste time listening to busy signals or to lengthy voice messages to learn that someone is on the phone or out of the office. Because users place phone calls simply by dragging the contact name to the toolbar, they never need to look up a number, and misdials are impossible.

The COM2001.com strategy has been to harvest, leverage and consolidate the most robust and cost-effective technologies available. It has converged PC-based communications servers, IP telephony, unified messaging, improved user interface design, and open architectures and industry standard protocols—and supports them with the most popular and powerful off-the-shelf software (Microsoft) and hardware (Intel and Dell) technologies.

The result is the InternetPBX, a unified communications solution that provides the most advanced, flexible and cost-efficient phone and messaging services for businesses today.

This Insider Perspective was contributed by David Perez—President and Co-Founder of COM2001.com. Our sincere thanks to Marc Pinotti, VP of Marketing at COM2001.com for his assistance with this COM2001.com overview.

5.4 Interactive Intelligence Enterprise Interaction Center

Interactive Intelligence, Inc.'s Enterprise Interaction Center (EIC)—is an all-in-one Windows NT-based Communication Server. EIC is an open, software-based unified solution that replaces a whole slew of proprietary devices such as PBXs, ACDs, IVRs, voice mail systems, and fax servers. Furthermore, Interactive Intelligence, Inc. offers additional products for e-mail response management, predictive dialing, and SS7-based pre-call routing.

5.4.1 Target Markets

Interactive Intelligence, Inc.'s three primary markets are: enterprises, call centers, and service providers. Enterprises would use EIC for departmental call center automation and unified messaging. Call centers would use EIC's sophisticated call routing, Web services, and multi-site communications integration. Service providers would use EIC to offer large-scale, subscriber-based unified messaging, conferencing, one-number follow-me, and IP-based services.

5.4.2 Features

EIC handles a variety of incoming communication channels such as phone calls, e-mails, faxes, Web chat requests, VoIP calls and Web callbacks. These incoming communication channels then can be routed appropriately based on agent proficiency, desire to use, or other criteria. EIC provides a graphical application generator to easily customize everything from dial plans and fax-on-demand, to ACD and IVR scripts. The versatility of EIC extends to the types of CO interfaces it supports, including analog, T-1, ISDN PRI, EuroISDN, and E-1. EIC's unified messaging feature allows users to access their messages from a PC, a Web-browser, or a touch-tone telephone. The product also provides tools for system monitoring and feature management, and EIC's unified architecture allows true end-to-end reporting across media types.

EIC greatly reduces the number of systems required to handle communications for call centers, enterprises, and service providers. It is ideal for organizations with remote locations, those wishing to leverage the Internet, or any interaction intensive organization such as banks, retail shops, and insurance companies. And, because EIC is an open, software-based solution, it can be easily customized to adapt to changes in customer needs or business models. This product emulates a totally unified Communication Server.

EIC is not a shrink-wrapped solution. Interactive Intelligence, Inc. sells EIC through a global network of channel partners. The sales process begins with a needs assessment. Once it is decided that EIC is a fit with the organization, the appropriate configuration is determined. Because of the architecture, EIC can be purchased incrementally, allowing the client to add features as they grow. Additionally, EIC's open architecture allows for seamless integration with third-party applications such as CRM, Sales Force Automation, and Help Desk software.

5.4.3 Test Drive

Interactive Intelligence, Inc. uses EIC for its own communication needs. EIC actually powers the Interactive Intelligence, Inc. Web site. Through their Web site, you can explore how EIC works. To try EIC out, you sign up using a Web-based form, your registration is immediately intercepted by a sales representative, who will promptly place a follow-up call to see if you have any questions. In addition, the sales representative has all your information in front of him. This is a useful example of how a unified system greatly enhances customer service.

While testing EIC, keep in mind how other enterprises would handle a similar situation. For example, when you call your credit card company you are asked to enter your 18-digit card number to verify that you are a customer. When you get an agent, you are asked to repeat this information. The reason the agent doesn't have your information is because the IVR system is separate from the computer system, which is separate from the ACD system, etc. Imagine if you could enter the card number and be able to contact an agent who had all your information in front of him or her without having to repeat information the system already asked for? EIC is well suited for this purpose—and without the integration hassles of Computer Telephony Integration (CTI).

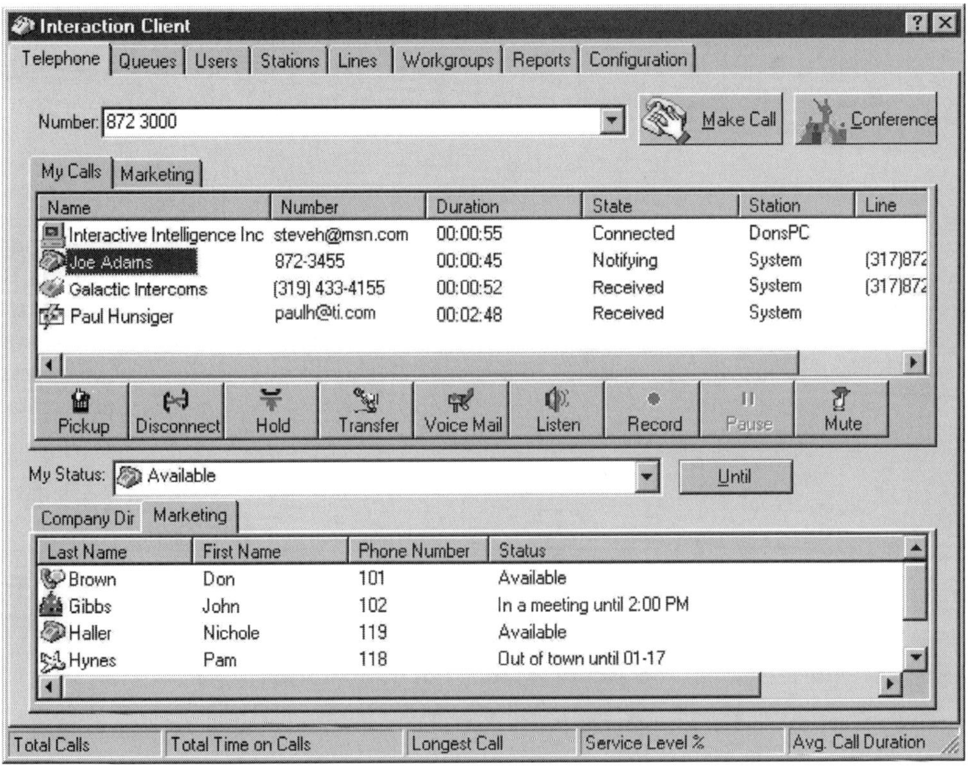

Figure 5-9: The Interaction Client is the portion of EIC that runs on each user's desktop PC. The interface is used to monitor and manage interactions of all types, such as telephony functions, Web chats, availability status and, to configure settings. Copyright (c) Interactive Intelligence, Inc.

Back at the Interactive Intelligence, Inc. Web-site, you have the option of contacting a sales representative either via e-mail, phone, callback, VoIP, fax or chat — choose chat. A small applet loads a chat client and instantly a sales rep initiates a chat session with you. You are able to ask questions in real-time, and he can even offer to bring in another representative to answer other, more specific, technical questions. Ask him to demonstrate some form of Web collaboration and he can, for example, make your browser open a local weather Web page. This feature could be used to guide customers to specific Web pages, for example, for a cross-sale.

On the Interactive Intelligence, Inc. Web-site, one can also interact with a specific company employee. You can search for a contact and the search result page will give you the option of initiating communications. You could initiate a callback, a chat session, a VoIP call, or send an e-mail. One great feature is a display showing the availability status of the requested contact (e.g. "On the Phone", "At Lunch", "Gone Home"). The employee can set availability status, and status options which are totally customizable. By showing an employee's availability status, the customer can then use the most convenient method to contact the employee.

5.4.4 Summary

The benefits of using a single platform to process all media types are well demonstrated by Interactive Intelligence, Inc.'s EIC product. A unified system reduces start-up costs, simplifies maintenance and administration, and gives a customer the flexibility to interact with an organization using the media type most convenient for her. Unification also allows for more efficient use of customer information, and faster and 'richer' responses to requests. If you want to see the future of communication servers, look no farther than Interactive Intelligence, Inc.'s EIC.

INSIDER PERSPECTIVE

PC-based Phone Systems:
The Multi-channel Interaction Management Platform

By Christine J. Holley—Manager, Market Communications,
Interactive Intelligence, Inc.

What is a PC-based phone system? Just try posing this question to vendors and you'll likely get a different response from every company that you ask. That's because the definition has been used to categorize so many non-traditional phone systems despite important differences in architecture and features. In addition, some of these "phone systems" are evolving so quickly that a definition one year ago, while accurate then, might not apply today.

One thing is for sure, there are as many variations on the PC-based phone system as there are flavors of Baskin Robbins ice cream. Some have transmuted to such a degree that one could argue they no longer even fit within what has become the "traditional" PC-based phone system market. One system in particular has redefined this market and, going a step further, has literally rebuilt communications from the ground up. This system, referred to as a multi-channel interaction management platform™, was developed by Interactive Intelligence® in 1994 and first released in 1997.

What's A Multi-channel Interaction Management Platform?

The impetus behind this unique architecture was prompted when the founder of Interactive Intelligence, Dr. Donald E. Brown, became intrigued at the loss of communication efficiencies as organizations grew and began adding multiple channels to their existing communication systems. These inefficiencies impeded productivity and frustrated customers. Dr. Brown viewed the classic approach to this problem as ineffectual. The old approach, that of deploying separate proprietary systems (for example, PBXs and fax servers) to handle different channels of customer contact, only exacerbated the problem by creating "islands" of communication and making it next to impossible to provide consistent service levels across media types.

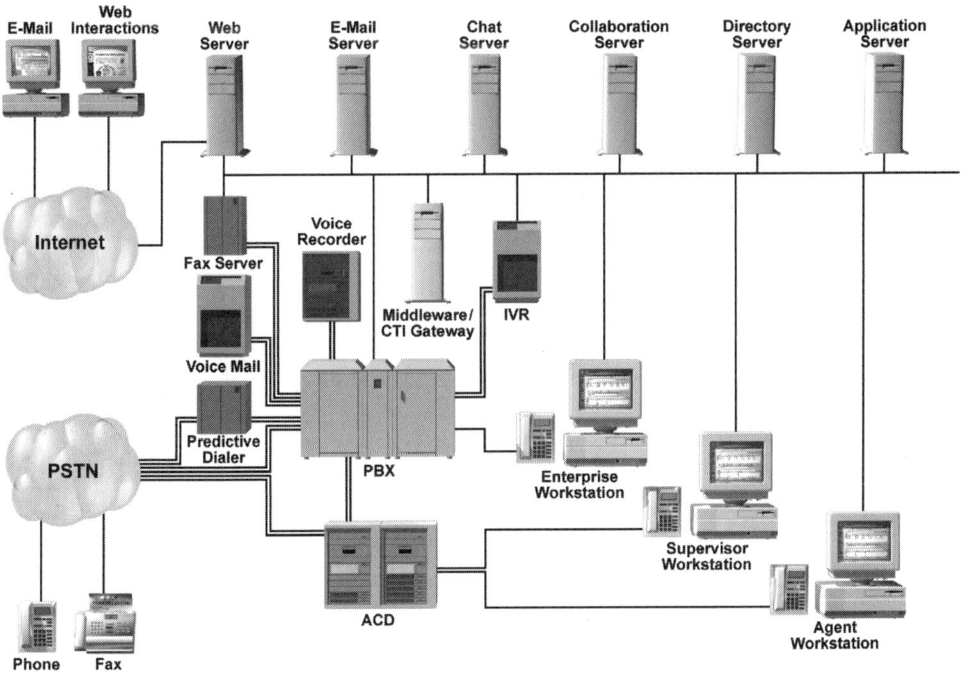

Traditional telecommunications architecture

Interactive Intelligence pioneered a new approach to communications by developing a software-based, multi-channel interaction management solution called the Enterprise Interaction Center(r) (EIC). This system not only differs from proprietary solutions, but is also distinguishable from the typical PC-based phone system in that it is capable of processing not only phone calls, but faxes, e-mails, and even Internet interactions. In fact, EIC was designed to replace PBXs, ACDs, IVRs, voice mail systems, fax servers, voice loggers, Web gateways, CTI middleware systems, and a variety of other communication devices.

At the center of this system's unique architecture is the Interaction Engine, a Java-based engine capable of processing hundreds of simultaneous events, and tens or even hundreds of thousands of events per hour. This Interaction Engine serves as a centralized point of control for any interaction type—from phone calls and faxes, to e-mail and Internet interactions. It allows for fewer points of failure, vastly simplified maintenance and administration, and true end-to-end reporting. EIC's comprehensive feature set and flexible architecture is ideal for several market segments including contact centers, enterprises, e-businesses, and service providers.

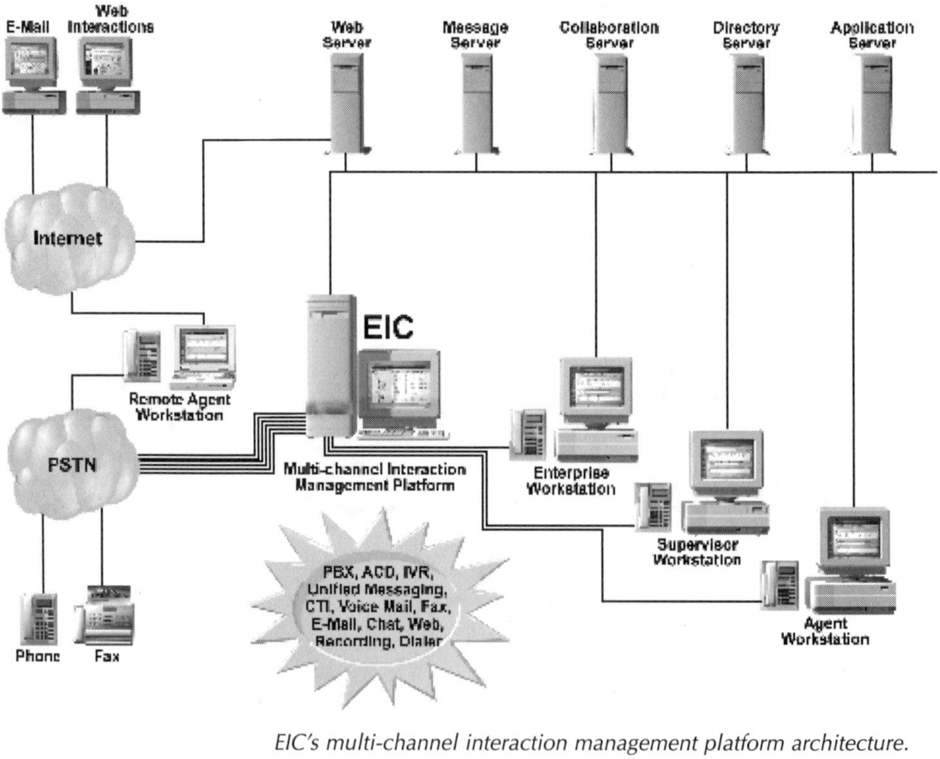

EIC's multi-channel interaction management platform architecture.

How Is This Platform Different From Network-based Solutions?

Since entering the communications market with its interaction management platform, Interactive Intelligence has witnessed a multitude of new players, including network-based vendors such as Cisco and 3Com. These vendors have elected to focus on the communication switching fabric itself. These solutions, while offering a breakthrough IP alternative to circuit-based phone systems, differ significantly from an interaction management platform solution. The primary difference is that the interaction management solution was designed as a multi-application platform supporting both circuit and IP-switched interactions. The whole premise of this architecture rests on the fact that end-users want a solution that isn't limited to a particular switching fabric or communication channel—they want a suite of communication applications that can support different switch protocols and new communication channels.

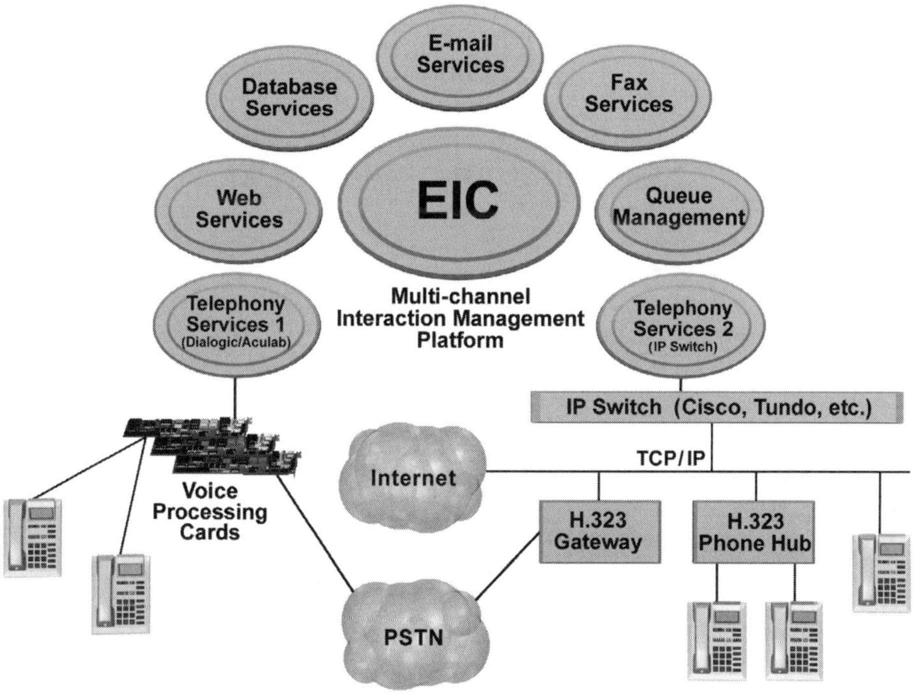

EIC's multiple telephony services architecture

EIC's interaction management platform does this by encapsulating its telephony functions into one component called Telephony Services (TS). TS allow the Interaction Engine to detect telephony events (for instance, incoming calls, DTMF digits, and call disconnects) and to perform operations on telephone calls such as transfer, conference, record, and others. Multiple versions of this component support either circuit- or IP-based switches. Thus, the circuit-based version could support Dialogic and Aculab voice processing boards, while its IP-based version could support a Tundo IP switch or a switch from Cisco. This architecture gives end-users the freedom to choose whatever 'flavor' of TS is right for them. Perhaps the most compelling part of EIC's architecture is that it offers this freedom while requiring no modification of its communication applications such as ACD, IVR, and unified messaging.

What's the Future of the Multi-channel Interaction Management Platform?

Patterned after the evolutionary law of survival of the fittest, the interaction management platform can support virtually any new technology—whether it comes in the form of a switching fabric, communication channel, or third party application. This ability to adapt and change protects the interaction management platform from the fate of obsolescence so common among new technologies today.

Interactive Intelligence will continue to develop this platform by incorporating the best that new technologies have to offer, and by partnering with leading vendors to offer complimentary solutions. Whatever form communication solutions take in future, you can be sure that the Multi-channel Interaction Management Platform will be the foundation on which they are based.

This Insider Perspective was contributed by Christine J. Holley—Manager, Market Communications, Interactive Intelligence, Inc. We would also like to thank Don Brown, President of Interactive Intelligence, for building one of the most responsive and conscientious teams in the industry—as well as one of the most exciting products in the industry.

5.5 InVADE International

UK-based InVADE International is the only European vendor covered in this PC-PBX market round-up. This Windows NT-based PC-PBX stands out for its unique architectural approach and the approach that InVADE has taken in marketing the product. Unlike the other PC-PBXs discussed in this overview, InVADE is designed primarily as a programmable platform for custom, mission-critical telephony applications, rather than as a turnkey solution. (However, some of InVADE's channel partners have developed turnkey solutions on the InVADE platform). Accordingly, InVADE is more often found in next generation call centers, than in small offices interested in out-of-the-box solutions.

The InVADE server is a standards-based platform for telephony applications, including ACD, IVR, voice over IP gateways, and more. These applications can be created with standard programming environments such as Microsoft Visual C++, Microsoft Visual Basic, Borland Delphi, and others. The multi-vendor applications built on the InVADE platform share access to the underlying telephony hardware resources. Additionally, the InVADE architecture is designed to support telephony hardware from multiple vendors. This architecture allows systems integrators to mix and match telephony hardware and software to suit their application needs.

InVADE has built a platform that is similar in its design goal to Dialogic's ECTF-compliant CTMedia platform; however the InVADE platform was developed within the Microsoft TAPI framework, and not the ECTF framework. The heart of the InVADE system is a TAPI service provider that provides call control, switch control, and call distribution capability.

Since InVADE is primarily designed as a platform and not a turnkey system, it is mainly marketed to systems integrators, IT departments looking to build custom telephony applications, and service bureaus looking for a flexible and scalable telephony platform.

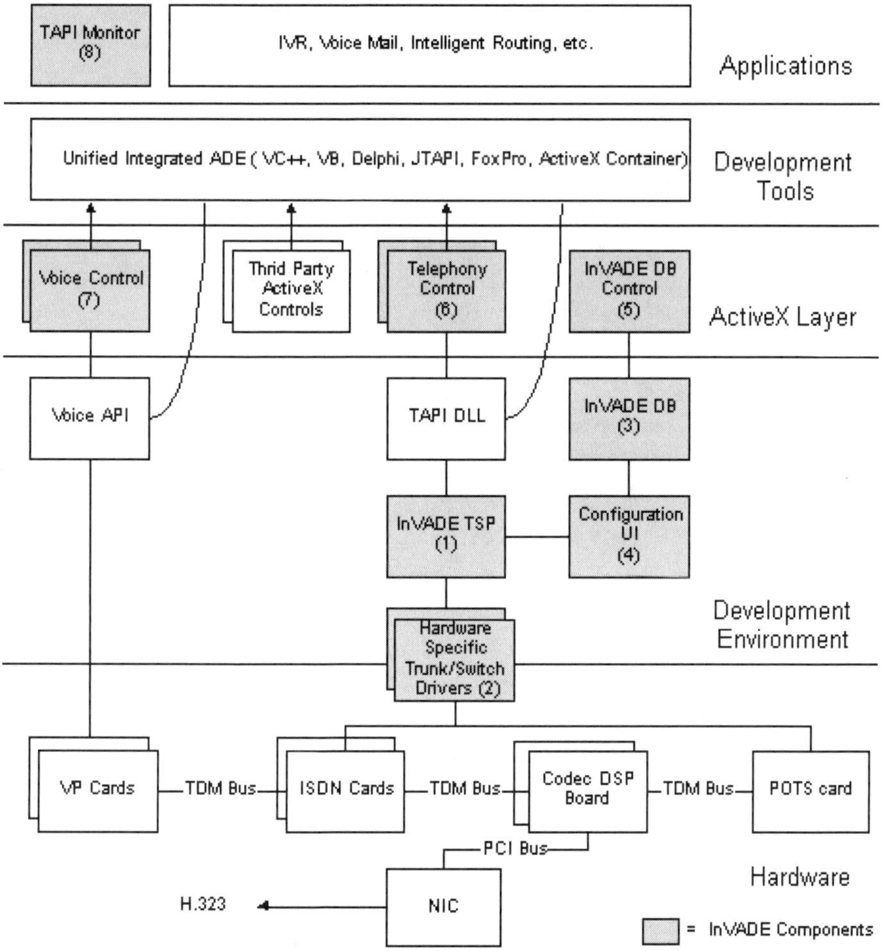

Figure 5-10: The InVADE architecture (Note: shaded boxes indicate InVADE components; other components are from third parties). Copyright (c) InVADE International.

5.5.1 InVADE Architecture

The discussion below, courtesy of InVADE, provides an overview of the InVADE architecture and all of the components that make up the InVADE solution.

- Telephony Service Provider (TSP)—The TSP (1) models Private Branch Exchange (PBX) functionality in a single software component, providing call control, switch control and call distribution capability. This is achieved by defining virtual devices, whose properties can be altered in real time, establishing correspondences between virtual and real devices and then managing communication between the virtual devices in accordance with server-based, user-defined rules.

  It is important to stress that InVADE devices are not representations of or interfaces to pre-existing physical devices, as is the case in modem-based or CTI link systems. InVADE can function as a true virtual PBX.

  The line devices defined by the TSP can function either as single lines or as line groups. These can be linked, permanently or dynamically, to physical devices on the bus (including, for example, voice resources) or to remote devices through trunk connections or wide area network gateways. These devices are then accessible through the Microsoft Telephony API (TAPI 2.0/2.1) or through TAPI-compliant applications. InVADE line devices support most common PBX call control functions, including make call, receive call, transfer, redirect, forward and conference.

  Line group devices can directly provide automatic call distribution (ACD) features, redirecting calls according to user-selected algorithms such as longest idle, circular and ordered. Physical resources can be allocated to a queued call without removing the call from the queue, meaning, for example, that a message can be played to the caller without disrupting the queue order.

  As the InVADE TSP is so closely coupled to TAPI, InVADE's line devices also support the configuration of TAPI features such as media modes, bearer modes and terminals.

  The InVADE TSP is installed as a standard Telephony Service Component and runs in the address space of Windows NT 4.0's Telephony Service Executable (TAPISVR.EXE)

  Line devices can be added and configured permanently using the Telephony object in the Windows NT Control Panel or dynamically by applications utilising the InVADE database control.

- Hardware Specific Drivers—The virtual devices meet reality through InVADE's Trunk Driver (2), which is a digital telephony call control interface to telephony hardware. As far as the user or developer is concerned, InVADE itself is hardware-independent, but this is of course achieved by the provision of hardware- or resource-specific drivers which enable the TSP to control telephony and TDM bus hardware. Currently available modules support ISDN digital interface cards, codec compression cards and the H.323 stack and POTS station interface cards. The TSP dynamically maps lines to physical and digital resources as required.

The Trunk Driver also controls the physical switching between resources on the TDM bus, conferencing and call progress tone detection.

When coupled to switching and telephony hardware an InVADE telephony line retains one crucial difference to those on other telephony and switching platforms: InVADE's lines are loosely coupled to the underlying physical devices. The association between the virtual lines and the hardware resources can be changed programmatically. This allows for much greater flexibility in program design and more efficient resource usage.

- The InVADE Database—Configuration details for lines and trunks are held in the InVADE database, currently implemented in MS Jet (3). The configuration utility (4) provides a user interface for maintaining the telephony device database. The database can also be maintained through the ActiveX database control (5), a COM server that exposes functions that allow the dynamic maintenance of InVADE lines and trunks. It can be used in environments that support Automation type languages such as Microsoft Visual C++, Microsoft Visual Basic and VB Script, Borland Delphi, MVB Script, JAVA.

  Line parameters include, for example, conditional call forwarding, number of active calls allowed, inbound or outbound call barring, auto answer and auto drop on remote disconnect. Trunk parameters include number of inbound and outbound calls, sub-addressing, conditional call forwarding and prefix addressing.

- Telephony Control—InVADE Telephony Control (6), also known as the SimplePhoneCtrl, is a simple COM object, which exposes TAPI 2.1 functions and events through a COM interface. It supports dispatch and custom interfaces and can therefore be used by languages such as Microsoft Visual C++, Microsoft Visual Basic, Borland Delphi and Java.

- Voice Control—InVADE Voice Driver (7) is a simple COM object which exposes high level voice processing functions (play, record, DTMF) through a COM interface. It supports dispatch and custom interfaces and can therefore be used by languages such as Microsoft Visual C++, Microsoft Visual Basic, Borland Delphi and Java. InVADE currently supports modular DSP boards (such as Pika Premiere and Aculab Prosody) which provide, on a single resource card: matrix conferencing, record and playback, tone detection, speech recognition, fax, and text-to-speech.

- TAPI Monitor—The InVADE TAPI Monitor (8) records all line and call activity, writing to a database using ADO (COM-based data access components). Real time activity is written to small, rapid-query tables, while historical data is written to larger tables for statistical analysis.

5.5.2 Summary

If you are interested in a next generation phone system platform that you can exploit for custom application development, then you should seriously consider InVADE. Of all the systems that we've looked at, InVADE is the most tightly integrated with the vision and architecture of Microsoft TAPI.

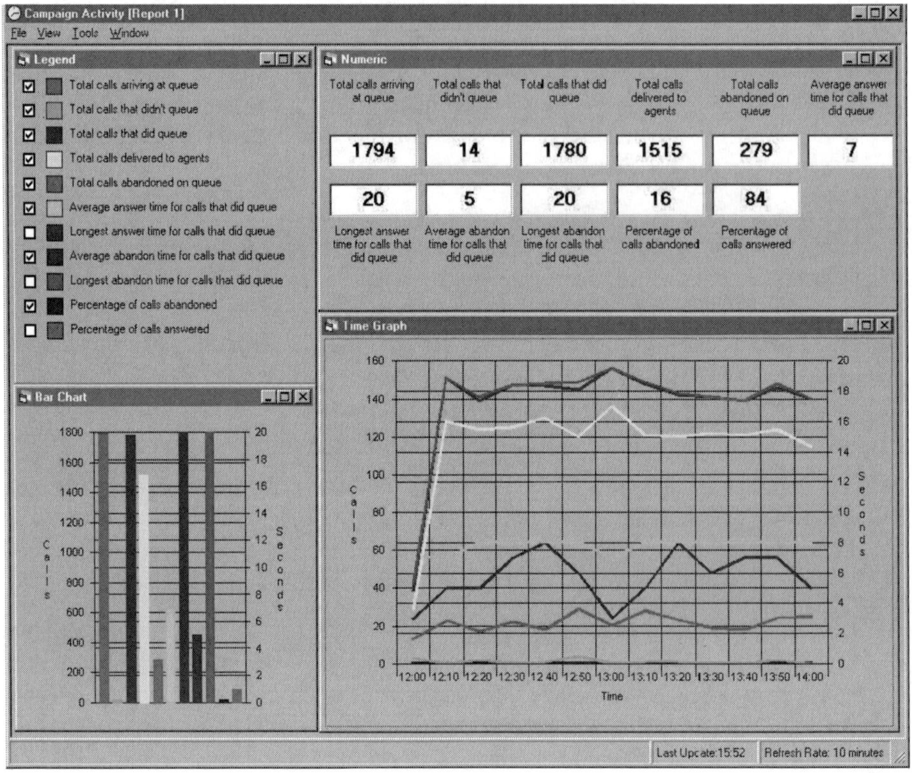

Figure 5-11: Output from the InVADE reporting module which tracks call center activity. Copyright (c) InVADE International.

INSIDER PERSPECTIVE

InVADE Overview and Benefits

By Claire Bunting—Technical New Product Development Director, InVADE International

InVADE International's channels provide the market with PC-PBX's with all the features of the traditional legacy platforms and more; open NT Communications Server to CTI-enable software used in a variety of Customers Contact strategies; and IP Gateways with intrinsic PBX/ACD features where support for voice and data convergence is paramount. In this very brief overview of the technology we outline the factors that we believe have been important in their selection of InVADE as the core technology underpinning their next generation solutions.

The Product

InVADE provides what is effectively an operating system for telephony and computer telephony hardware. The InVADE software delivers the core intelligence of a richly configurable and programmable communications platform. The software is a set of integrated modules that control all varieties of call control and media processing hardware, be they PC expansion boards or actual legacy PBX equipment.

Using InVADE alone, it is possible to build or simply configure an enormous range and scale of different types of telephony solutions. These may range from a simple 'audiotext' platform (e.g. 'dialup your Horoscope'), an IVR services platform (e.g. 'get your bank balance and list of most recent transactions'), or a standalone PBX system, through to a call center server with ACD, Voice Processing, VoIP and legacy PBX integration.

At the developer level, the InVADE software exposes its' rich functionality via ActiveX and TAPI interfaces. This allows solutions providers to easily extend the InVADE feature set and integrate with the core business processes.

The USPs for Developers

InVADE's main advantage lies in its careful design and implementation. InVADE is designed by solutions developers for solutions developers, and it delivers a combination of benefits not seen in any of its competitors:

- Modular design giving it functional scalability and hardware independence.
- Object oriented view of the call control and PBX functionality.
- Integration to the world's most common and well supported development environments (Visual Basic, Visual C++, Java, Delphi, etc.) by supporting TAPI and exposing its development components as ActiveX controls.

- TAPI direct, not through an integration layer, making it faster than those platforms with an indirect architecture. TAPI is a broader API than S.100 in terms of line and call control, is well supported and will probably dominate IP telephony at the client.
- "Feature On Demand" Resource Management.
- Transparent Voice over IP—for supporting enhanced Gateway to PBX interworking or Wide Area Distibuted Call Centers as deployed in the first such case of VoIP in a local Government establishment—see topology depicted below.
- Intrinsic ACD—InVADE supports ACD as an internal feature, and not as an "add on" application.

  As well as supporting an unmatched breadth of functionality, InVADE's benefits do not stop there. InVADE also has some unique features not seen in any of its competitors. Chief amongst these are:

- Direct Switching—allowing resources to be connected together without using, or being constrained by the PBX call control. This solves the biggest problem faced by CTI application developers.
- Custom configuration user Interface. InVADE provides access to its configuration data via an ActiveX control. This allows InVADE's PBX and ACD configuration to be re-presented in context.
- CEDC(—this is a unique, innovative, and cheap solution to "CTI-enabling" legacy PBX handsets. It can be extended to CTI enabling any handset, anywhere that can be connected to InVADE over a standard telephony network.

The Competitors

InVADE's design clearly contrasts with those of its competitors, whose architectures fall into one or more of the following four categories:

- Shrink-wrapped platform offering enhanced PBX features with integrated Voice and Fax applications and programmable options via proprietary development tools—like Interactive Intelligence's EIC and its Interaction Designer or Artisoft's TeleVantage and Visual Voice.
- Advanced, programmable PBX on NT that expose industry standard development interfaces through an extensible integration layer—like AltiGen's NetSwitch
- High Level Scripting tools or Application Development environments—like MediaSoft's IVS BlaBla or Visual Tools, or VB Voice from Pronexus.

- Feature rich API's—like CT Media from Dialogic, or ECTF S.100 based API's such as those available with Brooktrout's BOSTON architectured products.

What InVADE delivers encompasses all of these categories and more. While supporting Advanced PBX and a rich set of development features InVADE also provides a broad range of applications: ACD and PBX Configuration, MIS, Agent Desktop Phone, Operator/Supervisor Console, Voice Messaging, In-Q-Messaging. These are all provided free, and all with source code! InVA

DE is a TSP—the architecture is a true TAPI client-server implementation, which allows 'thin-client' applications to access telephony via the data network. It also lets you route calls to any TAPI enabled client terminals at remote centers over voice and Internet lines. For ISDN home workers a single connection can provide the same data and telephony access available in the office! InVADE has the added advantage that it will work with any TAPI enabled client PC 'virtual phone' software, contact managers or PIMs—leaving the user free to choose from a growing array of 3rd party offerings. Moreover, rather than a "proprietary" scripting language—the developer can use their preferred ADE such as VB, VC++, Delphi or Java etc.

Furthermore, unlike most of its competitors InVADE's design is fundamentally hardware independent (though we admit we have yet to support the breadth of manufacturers that we are capable of). This allows us to quickly support new and improved features (like VoIP and MC3) and should in the future allow us to benefit

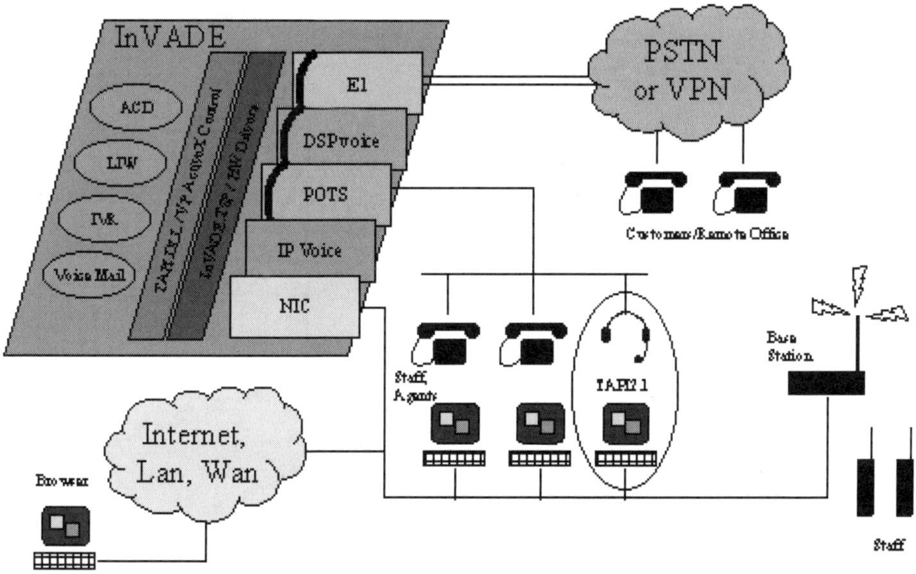

from alliances with many more hardware manufacturers. Current supported manufacturers are Aculab, Pika, Analogic, and Brooktrout with others planned this year.

Support for a variety of manufacturers via an industry standard TDM bus has allowed InVADE to provide a single platform support for scaleable IP Gateway solutions (from a single port BRI to multiple PRIs) with intrinsic support for enhanced PBX features and IVR applications via POTs or industry standard IP phones.

TAPI 3.0

InVADE is tightly bound to Windows NT Telephony Service and as such it implicitly exports its call control and line capabilities through Microsoft's Telephony API (TAPI). When InVADE first implemented its Telephony Service Provider (TSP) TAPI's use of terminals was heavily relied upon to enable programmable low-level resource switching. InVADE feels that this choice was vindicated as TAPI 3.0 has extended its use of terminals by adding more intelligence to them, and extending them with media stream control interfaces. This leaves InVADE well positioned to expose its current and new media stream capabilities via standard, integrated and well- supported interfaces. Tight support for the new features in TAPI 3.0 should allow InVADE to take better advantage of the growth of Voice over IP and internet media streaming to which TAPI 3.0 is clearly directed.

A Comparison of InVADE, CT Media, and Traditional PBXs

InVADE

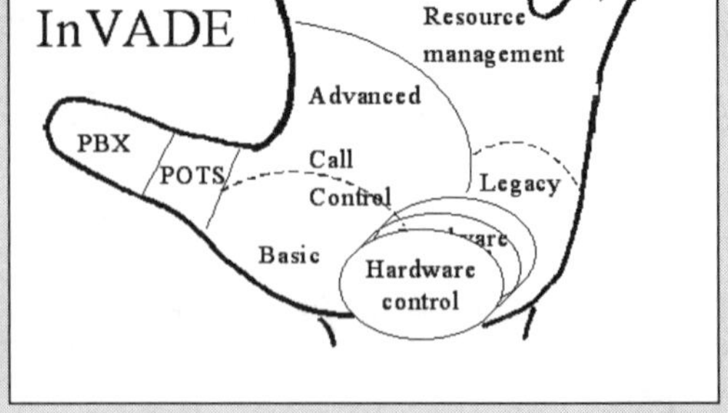

Core functionality—InVADE supports the delivery of calls between trunks, POTS and media processing resources. Call control features are extensive and InVADE has intrinsic support for ordinary groups and ACD groups and phantoms. InVADE has extensive support for media processing and resource management. InVADE's "middleware" architecture delivers hardware independence for both telephony network interfaces and media processing resources. As such, InVADE is able to support IP telephony within its core functionality.

Extensibility—InVADE provides a full call control, voice processing, conferencing, resource management and configuration management tool set. As InVADE runs within Windows Telephony service it supports TAPI for call control. Call Control and the other component features can be accessed via ActiveX controls. Media processing other than voice processing and conferencing can be easily integrated with InVADE's call control and resource manager. InVADE will soon support access to the current media types, and additional media types via TAPI 3.0

CT Media

Core functionality—CT Media API supports the delivery of calls to media processing resources. While the CT Media API has extensive support for media processing and resource management it provides only basic call control and no intrinsic PBX/handset support, although there may be supported PBX solutions available as applications from ECTF partners. The same is true for ACD and IP telephony.

Extensibility—The core API is S.100 (ECTF), although this standard API is supported by only a few manufacturers, and the CT Media implementation only supports Dialogic hardware. Solutions are implemented using "off the shelf" 3rd party applications or by writing new applications.

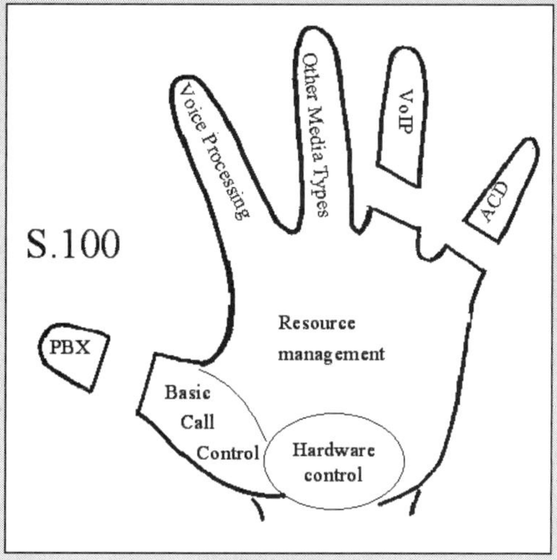

Traditional PBXs

Core functionality—The PBX supports the delivery of calls between trunks and POTS or Digital handsets. Call control features are extensive and the PBXs soft intelligence adds groups and phantom devices. Closely bound ACDs are usually available but tend to be expensive. Traditional PBXs contain no intrinsic media processing resources.

Extensibility—PBXs software and hardware tend to be proprietary and therefore CTI tends to be indirect, expensive and attempts to standardise CTI interfaces have had only limited success.

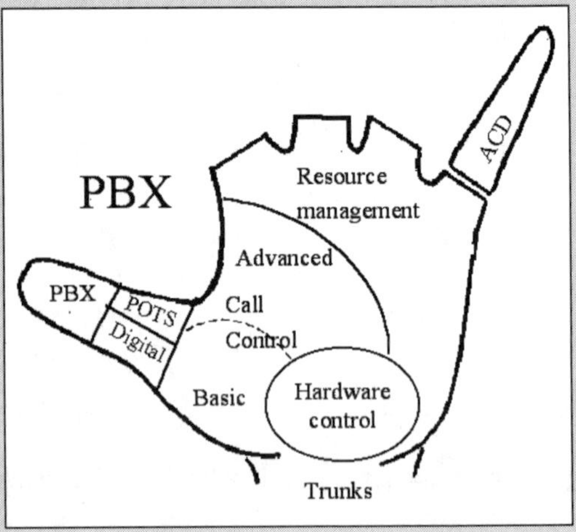

This Insider Perspective was contributed by Claire Bunting—Technical New Product Development Director, InVADE. We would also like to express our sincere thanks to Tom Thompson, the main architect and Technical Research & Development Director of InVADE. Claire and Tom are amongst the most insightful telephony architects in the industry.

Chapter 6

The Communication Server

6.1 Introduction

Communication Servers are designed to consolidate access and control of multiple media types in a single physical device. Today, most of these systems unify voice and data applications in a turnkey hardware appliance. The first Communication Servers begin appearing on the market in the late 1990s, when vendors such as UK-based Network Alchemy introduced its ArgentOffice product line. Over the past few years, an increasing number of vendors have introduced Communication Servers. Most of these vendors are targeting the small office or branch office market where scalability requirements are limited and it is more efficient to configure and administer a single communication device, rather than multiple independent systems.

In general, Communication Server vendors promote three main benefits:

1. The cost savings achieved by consolidating multiple communication devices into one and reducing the number of phone lines and/or service providers (since the same phone lines/service providers can be shared between voice and data).

2. The employee productivity benefits gained by enhanced communication applications such as unified messaging, CTI, and remote access solutions.

3. The reduced administrative costs by eliminating the need for specialized staff or high-priced consultants for ongoing system administration and configuration.

In this chapter, we'll take a close look at Communication Servers, including trends, standards, and products.

6.2 Market Trends

The first category of next generation phone systems that we explored was the PC-PBX. PC-PBXs integrate voice communication with the programmable, standards-based environment of the PC. Communication Servers, on the other hand, integrate voice communication with the turnkey modularity of data network access devices, such as network routers and hubs. (In fact, a number of the companies that develop Communication Servers were previously involved in the development of network devices). Accordingly, the evolution of Communication Servers has followed a different path than PC-PBXs. The main trends influencing the Communication Server market today are discussed below.

- Communications Servers typically connect to high-bandwidth digital phone lines, such as ISDN, T-1/ E-1, or DSL. Many Communication Servers provide voice and data access through a single trunk interface. For example, ISDN-based solutions will dynamically allocate bandwidth between data and voice applications as necessary—setting up or tearing down channels at the appropriate time. Recently, a number of vendors began offering solutions that support voice over DSL (see *The Trend Toward DSL* below).

Cultural Differences

The standards for digital line interfaces vary from country to country. As a result you will often find that a Communication Server popular in one country is unknown in another. For example, UK-based Network Alchemy (a subsidiary of Lucent Technologies) has developed an ISDN-based product line since ISDN is the prevalent digital line standard in the UK and much of Europe. Network Alchemy has had phenomenal success in Europe, but is relatively unknown in the US. In the US, ISDN is being quickly replaced by newer technologies such as DSL, and as a result, ISDN-based products are not receiving much market attention. In fact, at least one US company, Jetstream Communications, has replaced their ISDN-based product line with a DSL-based product line. Communication Server vendors design their products to "speak" the digital interface language of their target market.

- Support for real-time media transmission over packet-switched networks is becoming increasingly common. Just as with PC-PBXs, this feature is often introduced first as an expansion bus interface that allows multiple Communication Servers to be linked over a LAN or WAN. In addition, some products also support H.323 compatible VoIP to the desktop, allowing users to handle voice calls from PC applications such as Microsoft Windows NetMeeting 3.x or standard desktop IP phones.

- Web-based configuration and administration are standard. Almost all of the Communication Servers that we've seen provide a Web browser interface for both system administration and individual user account administration. Web-based administration interfaces make Communication Servers ideal for branch office installations where most of the administration happens remotely.
- Scalability is typically limited to the branch office or small to medium enterprise (SME) market. Today, most Communication Servers will support up to a few hundred users. Accordingly, most vendors are targeting the branch office or SME market. However, with VoIP-based expansion bus interfaces, many vendors are looking to provide scalability to the enterprise level.

The Trend Toward DSL: Integrating Voice and Data over DSL

By Steve Gleave—VP of Marketing, Jetstream Communications

DSL broadband access networks can now deliver high-speed Internet access and local dial tone services to the small and mid-size business markets.

The Digital Subscriber Line (DSL) phenomenon is well known to those involved in communications and networking. The technology permits the data carrying capacity of the installed base of over 800 million copper twisted wire pairs to be expanded.

To date, this expansion has been targeted at data communication applications, such as high-speed Internet access and remote LAN connectivity. As a data transport service, DSL is well positioned between inexpensive analog dial-up service and expensive dedicated T-1 service. The transmission speeds it provides, the continuous nature of its connection and the price at which it can be delivered make DSL service an ideal data solution for most small and mid-size businesses. Indeed, a number of Incumbent Local Exchange Carriers (ILECs) and Competitive Local Exchange Carriers (CLECs) are now busy deploying DSL broadband access networks.

With the upcoming availability of Voice over DSL (VoDSL) broadband access solutions, customers can now step back and look at DSL from a somewhat broader perspective. By enabling the delivery of up to 16 telephone lines and high-speed Internet access over a single DSL connection, VoDSL may be the technology that will allow the Small and Medium Enterprise to purchase a compelling bundled service offering from a range of new or existing local telephone companies.

The voice quality and reliability of the telephone service delivered by VoDSL solutions are comparable to that provided by conventional telephone service. In addition, these solutions support all currently available custom calling features, such as caller identification, call waiting and call forwarding, and the telephone service can be used with the subscriber's existing equipment, including individual telephones, telephone systems, fax machines and modems. For high-speed data transmission,

VoDSL solutions allow subscribers to connect their local data networks directly to service providers' high-speed data networks. Subscribers gain the benefit of dynamic bandwidth allocation, which allows unused voice traffic capacity to be used for high-speed data access.

VoDSL broadband access solutions are now hitting the market, although voice services delivered over DSL may not be marketed as such. However, you may find that VoDSL solutions are marketed as service bundles, whereby your telecommunications carrier will offer you a given number of voice lines (typically between 8 and 16), a contracted number of local and long distance minutes, plus a full range of enhanced telephony services (call waiting, forwarding, voice mail etc.) and an always-on Internet service.

VoDSL—How it Works

In general, a VoDSL system functions as an overlay solution to a DSL broadband access network, enabling a telecommunications service provider (TSP) to extend multi-line local telephone service off of a centralized voice switch. For example, Jetstream's VoDSL solution allows up to 16 telephone lines and high-speed continuous data service to be provided over a single DSL connection.

A VoDSL solution typically consists of three components (see below). First, a carrier-class voice gateway that resides in the TSP's regional switching center (RSC) and serves as a bridge between the circuit-based voice switch and the packet-based DSL access network. The second component, an Integrated Access Device (IAD), resides at each subscriber premise and connects to a DSL circuit. It also serves as a circuit/packet gateway and provides the subscriber with standard telephone service via up to 16 analog POTS ports and Internet service via an Ethernet connection. The last component is the management system—owned and operated by the TSP.

With VoDSL solutions, DSL broadband access networks now have the coverage, capacity and cost attributes to enable TSPs to deliver local telephone services as well as data services to the small and mid-size business markets. Considering the rate at which TSPs are now embracing VoDSL solutions, chances are that the service will be available to you in the not too distant future. Details on TSPs who are rolling out VoDSL solutions may be found on the Web sites of VoDSL vendors, including www.jetstream.com.

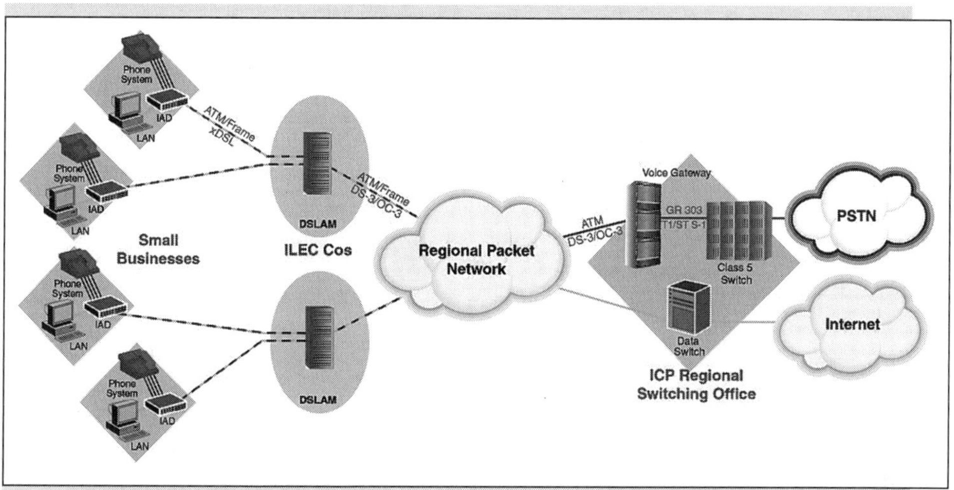

Sample VoDSL Access Solution

It has already been established that DSL access networks have the right bandwidth to serve the data needs of small and mid-size businesses. With VoDSL access solutions, this is true for serving the local telephone service needs of those subscribers as well. As mentioned above, some VoDSL solutions are capable of delivering 16 telephone lines over a DSL circuit along with standard data traffic. Since ninety-five percent of small businesses use twelve or fewer telephone lines, a single DSL circuit provides sufficient bandwidth to serve the voice needs of the vast majority of the market. In addition, if more than 16 lines are required, most VoDSL solutions allow a TSP to scale service by provisioning additional DSL connections.

In addition to providing the right capacity for providing local telephone service, DSL broadband access networks are very efficient in the way they deliver service. TDM-based transport services, such as a T-1 line, require the bandwidth of the line to be channelized and portions dedicated to certain services, such as a telephone line. Even if a call is not active on that line, the bandwidth allocated to that line cannot be used for other purposes. DSL access networks are packet-based, allowing VoDSL solutions to use the bandwidth of a DSL connection dynamically. VoDSL solutions only consume bandwidth on a DSL connection when a call is active on a line. If a call is not active, then that bandwidth is available for other services, such as Internet access. This dynamic bandwidth usage allows TSPs to maximize the potential of each DSL connection, delivering to subscribers the greatest number of telephone lines and highest possible data speeds.

Because telephony traffic is more sensitive to latency than data traffic, VoDSL solutions guarantee the quality of telephone service by giving telephony packets priority over data packets onto a DSL connection. (ATM is typically used as the transport layer for VoDSL solutions in order that this level of prioritization may be assured). In other words, telephony traffic always receives the bandwidth it requires and data traffic uses the remaining bandwidth. Fortunately, telephony traffic tends to be very bursty over the course of a typical business day, so the average amount of bandwidth consumed is minimal. For example, over a single 768Kbps symmetric DSL connection, a TSP could provide eight telephone lines (serving a PBX/KTS with 32 extensions) and still deliver data service with an average speed of 550Kbps (see below).

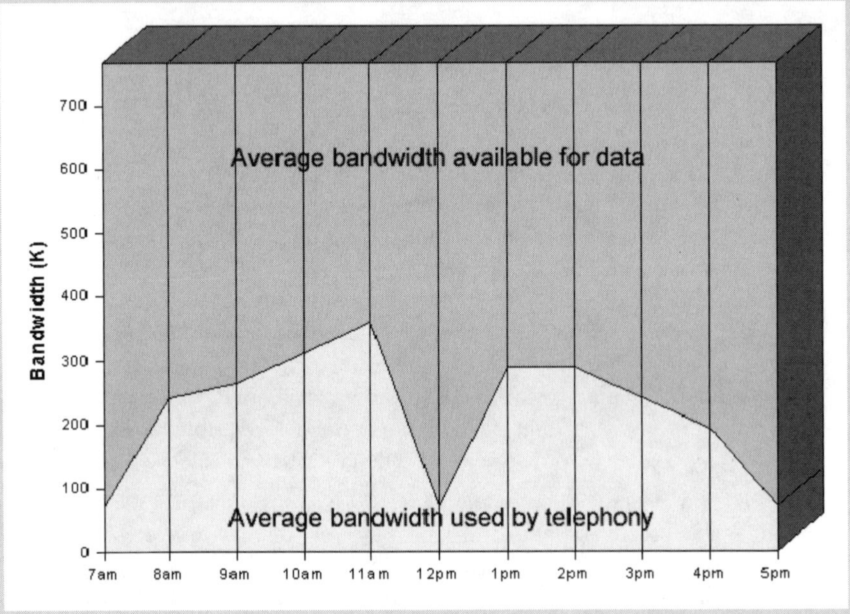

Bandwidth usage of a 768Kbps SDSL circuit (Assumptions: The 8 lines are servicing a 32 station PBX/KTS at a P.01 grade of service. 10-hour day with a 15% busy hour.)

Integrated Access Devices

In order to connect to VoDSL services, an Integrated Access Device (or IAD) is required at the customer premise. IADs connect the subscriber's network to the broadband access line. A TSP installs one or more IADs at each of their subscriber's premises to aggregate voice and data traffic onto a single broadband access line. The IADs are typically wall-mounted in a wiring closet and provide a subscriber up to 24 standard telephone lines as well as Internet and data service via a standard

Ethernet or V.35 local network connection. The IADs do not require the subscriber to deploy any new telephone (or data routing) equipment—existing telephones, fax machines, key systems or PBXs can all connect to the IAD. When new IADs are installed in a network, they can be automatically configured by the centrally located VoDSL gateway. This ensures that TSPs will be able to quickly add or delete telephone lines for the end user as required. For example, more telephone lines could be added for a seasonal calling campaign or special promotion.

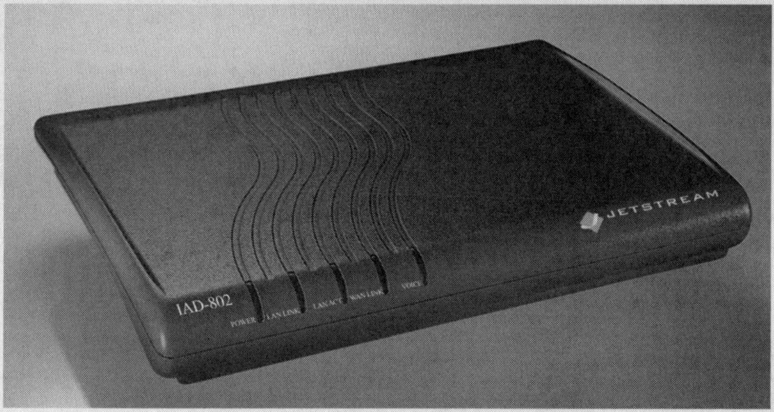

The Jetstream IAD-802 Integrated Access Device

The Future of VoDSL

VoDSL is the first technology through which telecommunication service providers can cost-effectively deliver a suite of integrated communications service bundles across broadband access media. Once broadband connectivity has been established between the customer premise and the service provider's network, the opportunity is presented for service providers to deliver an exciting range of next generation services to both business and residential customers. VoDSL gateways are highly flexible platforms that will support the creation and delivery of new broadband-based, value-added voice services. Examples of these new services may include advanced call handling, call processing, real-time call control and messaging functions, all of which could be offered by the service provider from within their telecommunications network. It is likely that these next generation services will be packaged as part of a service bundle, specifically targeted at a particular customer's business requirements. By integrating control for these next generation voice services into a familiar Web browser, customers will quickly derive the benefit of these enhanced applications through intuitive and familiar interfaces. The range and value of these next generation services is virtually unlimited. Indeed, the ability to create personalized services will even be extended out to the end users, thereby giving customers the freedom and flexibility to define their own particular service offerings.

VoDSL is laying the foundation for broadband service connectivity to the mass market. Once this foundation is in place, next generation services will quickly be deployed to help service providers differentiate their service offerings and provide compelling value-added solutions for their customers.

Our sincere thanks to Steve Gleave, VP of Marketing at Jetstream Communication, for this contribution. Steve has over 17 years of experience in the telecommunications industry, starting with GEC Telecommunications in the U.K. His wide-spanning roles have included technical, business development and marketing positions at Newbridge Networks throughout Europe, Africa and the Middle East, as well as the launch of Premisys' UK operations. He holds a 1st Class Honors Degree from Bristol University in Electrical and Electronic Engineering. Jetstream Communications is one of the leaders in Voice over DSL (VoDSL) infrastructure products for both carriers and small businesses.

6.3 Relevant Standards

With the PC-PBX, we saw that the relevant standards (the ECTF framework and TAPI) mainly describe the internal architecture of a PC-PBX. These standards are designed to provide customers with the ability to mix and match the telephony hardware and the software for a PC-integrated solution.

For the most part, Communication Servers are turnkey network appliances. The internal architecture of these systems is largely proprietary. Just like network routers or hubs (and traditional PBXs, for that matter), the hardware expansion modules that work with one vendor's system will not work with another's. Accordingly, the relevant standards in the Communication Server industry mainly specify the interoperability between Communication Server appliances themselves or between appliances and peripherals (such as IP-based phones). The standards that facilitate this type of interoperability specify call control and media processing over packet-switched networks. The most widely supported of these standards are ITU H.323, IETF SIP, and IETF MGCP.

6.3.1 ITU H.323, IETF SIP, and IETF MGCP

International Telecommunication Union (ITU) H.323, Internet Engineering Task Force (IETF) Session Initiation Protocol (SIP), and IETF Media Gateway Control Protocol (MGCP) are standards that define a call control and signaling infrastructure on IP-based networks. More specifically, these standards define how devices on an IP network establish, monitor, maintain, and teardown connections with one another. Although H.323 is the *de facto* standard, the other protocols are quickly emerging as viable alternatives.

Many Communications Servers use these standards as the signaling protocol in a distributed architecture, where an IP network provides a "virtual tie line" between multiple Communication Server units. Some systems also use these protocols to support desktop applications such as Microsoft Windows NetMeeting or desktop IP telephones.

Standards Tutorial: Session Initiation Protocol (SIP) Overview

By RADVision, Inc.

What is SIP?

The Session Initiation Protocol, or SIP, is a new IETF signaling protocol for establishing real-time calls and conferences over Internet Protocol networks. Each session may include different types of data such as audio and video, although currently most of the SIP extensions address audio communication. As a traditional text based Internet protocol, it resembles the hypertext transfer protocol (HTTP) and simple mail transfer protocol (SMTP). SIP uses Session Description Protocol (SDP) for media description.

SIP is independent of the packet layer. The protocol is an open standard and is scalable. It has been designed to be a general-purpose protocol. However, extensions to SIP are needed to make the protocol truly functional in terms of interoperability. SIP also enables personal mobility by providing the capability to reach a called party at a single, location-independent address.

Relationship between SIP and H.323

Both SIP and H.323 define mechanisms for call routing, call signaling, capabilities exchange, media control, and supplementary services. SIP is a new protocol but promises scalability, flexibility and ease of implementation when building complex systems. H.323 is an established protocol that has been widely used because of its manageability, reliability and interoperability with PSTN. There is a general consensus among standards organizations, companies and technology experts that standardized procedures need to be specified to allow seamless interworking between the two protocols. Bodies such as TIPHON (ETSI), aHIT! (IMTC) and IETF are working to address this topic.

RADVision is leading the ITU-T SG16 initiative to define SIP guidelines, and is working closely with other ITU members to achieve true end-to-end connectivity between the two protocols on the same network infrastructure providing global services to end-users. In order to achieve an interoperable system, the following needs to be addressed:

- Topology Definition: The architecture needs to be defined in terms of the respective entities. Call scenarios and interfaces must then be developed based on the defined topology.
- Supported Services: Advanced services such as conference calls and supplementary services should be incorporated.
- Supported "Data Capabilities": H.245 channels to Session Description Protocol (SDP) mapping are essential to respond to the issue of supporting audio and multimedia support.
- Protocol Versions: The protocol versions of both SIP and H.323 need to be addressed in order to define the scope of the work.

SIP Architecture

SIP's basic architecture is client / server in nature. The main entities in SIP are the User Agent, the SIP Proxy Server, the SIP Redirect Server and the Registrar.

The User Agents, or SIP endpoints, function as clients (UAC's) when initiating requests and as servers (UAS's) when responding to requests. User Agents communicate with other User Agents directly or via an intermediate server. The User Agent also stores and manages call states.

SIP intermediate servers have the capability to behave as proxy or redirect servers. SIP Proxy Servers forward requests from the User Agent to the next SIP server User Agent within the network and also retain information for billing/accounting purposes. SIP Redirect Servers respond to client requests and inform them of the requested server's address. Numerous hops can take place until reaching the final destination. SIP's tremendous flexibility allows the servers to contact external location servers in order to determine user or routing policies and therefore does not bind the user to only one scheme to locate users. In addition, in order to maintain scalability, the SIP servers can either maintain state information or forward requests in a stateless fashion.

The third entity comprising SIP is the SIP Registrar. The User Agent sends a registration message to the SIP Registrar and the Registrar stores the registration information in a location service via a non-SIP protocol. Once the information is stored, the Registrar sends the appropriate response back to the user agent.

Session Initiation

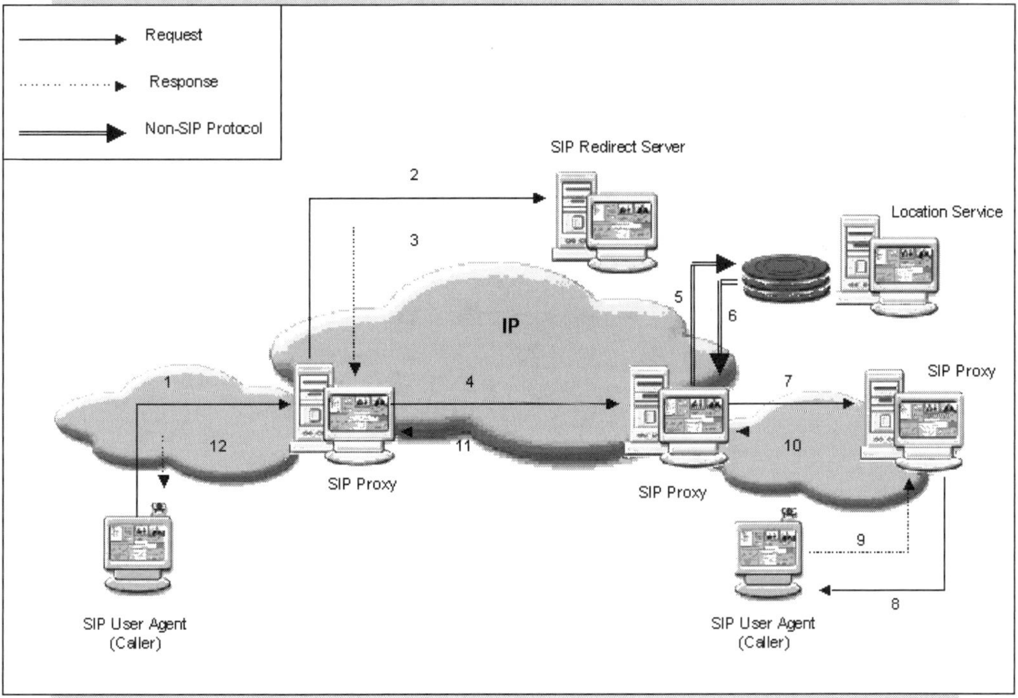

SIP Registration

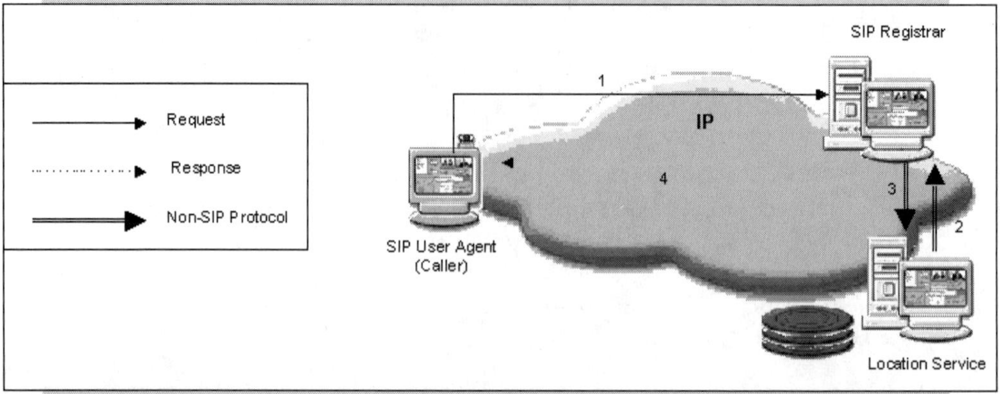

Our thanks to Karen Gurwitz of RADVision for this contribution.

CHAPTER 7

COMMUNICATION SERVER MARKET OVERVIEW

This chapter provides an overview of the leading Communication Servers on the market today, including Lucent Technologies Alchemy Series, Praxon PDX, and Vertical Networks Instant Office.

7.1 Lucent Technologies Alchemy Series

There are four products in the Lucent Technologies Alchemy Series portfolio: CyberGear TA, CyberGear Gold, ArgentOffice, and ArgentBranch. Both CyberGear TA and CyberGear Gold are targeted at individual power users or very small offices. Although these products are introduced below, the majority of this discussion will be devoted to the more scalable and feature-rich ArgentOffice and ArgentBranch platforms.

7.1.1 Overview

Alchemy ArgentOffice and ArgentBranch are scalable ISDN routers, LAN hubs, and PBXs designed for small and medium enterprises (SMEs) or branch offices of up to about 100 users. In addition, the ArgentBranch platform allows voice and data calls to be passed between ArgentBranch based systems over a LAN or WAN using voice over IP. This platform also provides an H.323-compliant IP telephony gateway, which is currently able to handle IP calls from Microsoft Windows NetMeeting software.

An ArgentOffice or ArgentBranch solution consists of two or more turnkey units attached by expansion cables. One of the units provides the ISDN network interface and voice/data routing support while the other unit provides the interface to analog telephones, fax machines, or modems. The analog interface units are available in 8 or 16 port configurations and are stackable to 96 ports (ArgentOffice) or 100 ports (ArgentBranch).

A basic system includes a dual ISDN BRI telephone network interface module. If more bandwidth is needed, this module can be replaced by a quad ISDN BRI interface or an ISDN Primary Rate Interface (PRI) module. ISDN PRI service provides support for up to 30 voice or data channels of 64 Kbps each. The same unit that provides the telephone network interface also includes an integrated 8-port Ethernet hub and a WAN interface port for leased line data connectivity.

The physical appearance of the ArgentOffice and ArgentBranch units is similar to that of a network router or hub and will be familiar to anyone with a datacom background. Even traditional telecom resellers and specialists will find the turnkey design easier to approach than an entirely computer based phone system that requires the installation of PC expansion cards. Additionally, it's possible for professionals from either the datacom or telecom background to ease into the world of converged voice and data by starting with the feature set that they are comfortable with while learning about the other.

Figure 7-1: Lucent Technologies ArgentOffice, a stackable voice/data Communication Server. Copyright (c) Lucent Technologies.

7.1.2 Test Drive

The initial setup of ArgentOffice is straightforward and does not consist of much more than plugging in phone lines and cables. You configure the ArgentOffice unit using a Windows application that runs on one of the PCs connected to the same LAN as the unit.

When running the configuration software for the first time a wizard is launched that prompts you for all of the essential system parameters. Each system configuration screen can also be accessed directly from a graphical toolbar that runs across the top of the configuration software screen.

The configuration software automatically locates the ArgentOffice unit on the LAN and requests a name for the unit, a password, an IP address, and a number of other basic parameters. The complete setup of the ArgentOffice system includes the configuration of the following features:

- Extensions—Each extension in the system is given an identifier, which is usually the name of the extension owner. This identifier is passed as Caller ID information on internal calls. You set the parameters for call forwarding and voice mail

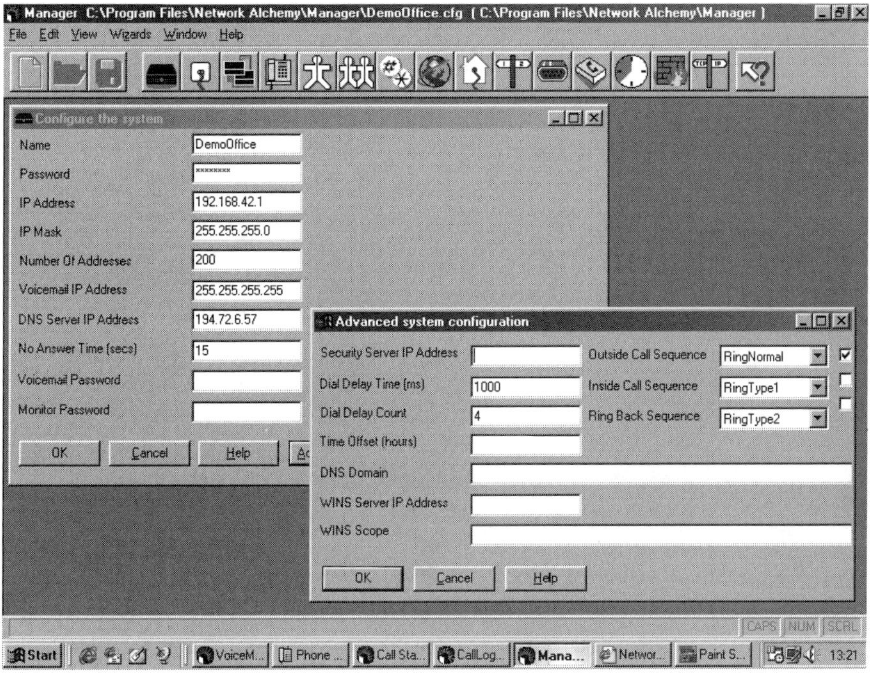

Figure 7-2: ArgentOffice configuration screen. Copyright (c) Lucent Technologies

for each extension using a simple graphical form. The extension configuration can also be performed by each extension owner using Alchemy's client software.

- Hunt Groups—What Alchemy calls Hunt Group Support is actually basic automated call distribution (ACD) support. Extensions are added to workgroups and workgroups are configured to distribute calls in a specified manner. ArgentOffice supports the following distribution schemes: round-robin, first available, longest idle, and broadcast. Alchemy Hunt Group Support is an extremely useful tool for informal call centers and workgroups.
- Data Service—Connections to an Internet service provider (ISP) for high-speed data access can be configured with a number of ISDN-specific parameters. After entering basic information, such as the ISP's name and phone number, you set parameters for controlling the dynamic allocation and deallocation of data channels. These parameters specify bandwidth thresholds that ArgentOffice uses to determine if it should set-up or tear-down data channels. For example, setting a bandwidth threshold to 50% will cause ArgentOffice to automatically allocate a new data channel if the bandwidth on a channel exceeds 50%. Likewise, thresholds can be set that will cause ArgentOffice to tear-down ISDN channels.
- Firewall Configuration—ArgentOffice includes a built-in firewall to prevent unau-

thorized access of your LAN. The firewall can be used to filter out specific Internet protocols, specific IP addresses, or even a specific user-defined data stream.

- Voice Mail—The ArgentOffice voice mail system resides on any PC connected to your LAN. All voice messages are stored as WAV files and can optionally be forwarded as attachments to any SMTP compatible e-mail system. Aside from providing an IP address for the server running the voice mail software, ArgentOffice requires very little additional configuration information.

The ArgentOffice configuration program also provides screens for setting up telephone shortcut codes, a "yellow pages" directory for Caller ID lookup on incoming calls, and much more. The system also includes complete call logging support, integration with frame relay services, least cost routing, call recording and monitoring, conferencing calling, music-on-hold, and much more.

On the ArgentOffice CD you will also find a basic desktop CTI application that can be installed on each user's Windows PC. This program initiates a "screen pop" to announce incoming calls and allows users to perform simple call control, such as forwarding calls or placing calls on hold. The application also notifies users of voice mail messages and allows users to configure their extensions. ArgentOffice users will find the busy lamp application useful for viewing the status of other extensions on the system.

Finally, ArgentOffice provides an operator's console to allow a system operator to view and manage hunt groups and extensions.

A Power User's Dream: CyberGear TA and CyberGear Gold

Across Europe, ISDN Basic Rate Interface (BRI) is the most common phone line interface for power users looking to upgrade from analog service. ISDN offers sub-second call setup, high-speed transmission, dynamic channel allocation, multi-channel operation, and Multiple Subscriber Numbering (MSN). All of the Alchemy products are designed specifically for EuroISDN and take full advantage of the ISDN feature set.

The Alchemy CyberGear TA and Alchemy CyberGear Gold each provide connectivity to a single ISDN BRI line. ISDN BRI provides 128 kbps of available bandwidth divided between two 64 kbps bearer (B) channels. ISDN BRI also includes one data (D) channel for call signaling information. The Alchemy CyberGear products are a perfect complement to the features of ISDN BRI service.

CyberGear TA is an ISDN BRI terminal adapter and three extension PBX designed primarily for a single user. The system provides high-speed Internet access to a single PC connected to the unit via a serial link. The three phone extensions can be used for standard analog telephones, fax machines, or modems. Using ISDN's MSN feature and Alchemy's support for Direct Dial In (DDI), callers can directly dial any of the devices connected to the three extensions.

Additionally, CyberGear TA provides support for Microsoft TAPI to allow contact management software and other telephony enabled Windows software applications to interact with the telephony features for the unit. For example, outbound calls can be initiated simply by clicking on a name in a contact manager database. All of the Alchemy products also pass Caller ID information through to the phone extensions.

As far as appearances are concerned, CyberGear Gold has exactly the same futuristic design and curved plastic case as CyberGear TA. However, with support for six phone extensions and a 10Base-T Ethernet connection, CyberGear Gold is more appropriate for a small office than an individual user. CyberGear Gold is configured with a Windows administrative interface and supports a suite of Windows productivity applications, including desktop call control, busy lamp, and more. The same software used with Alchemy ArgentOffice is also used with CyberGear Gold.

7.1.3 Summary

ISDN service requires an advanced communication platform to take full advantage of its feature set. Lucent Technologies Alchemy Series provides this platform and packages it in a way that is accessible to datacom professionals, telecom professionals, and skilled end-users.

INSIDER PERSPECTIVE

Birth of the Network Appliance

By Robin Hayman—Alchemy Sales & Marketing Manager, Lucent Technologies

The basis for the development, which led to the birth of the Network Appliance as typified by the ArgentOffice and ArgentBranch platforms, came from a typical SME business- our own! Scorpion Logic was a UK based hi-tech start-up (as they were called in the days before dot coms existed) developing entry-level ISDN routers for the SME & remote branch office marketplace. As such, we had a fair number of long-haired, open-toed sandal wearing, pizza munching, Jolt swilling rocket-scientists masquerading as the development team. In terms of PC networking, Internet access and general data connectivity, we obviously installed, managed and maintained our own network. But when it came to our telephone system (Panasonic KXTD) any adds, moves or changes required a call to our supplier to send a "man-in-a-van" to do it for us—frustrating or what! As a company, life was good for us and through acquisition we ended-up working for Bay Networks (now Nortel Networks) developing the Nautica product line (Clam, Marlin, etc.). At that point we were a small remote office of a large multi-national operational, with a different telephone system (Lucent Definity as it so happened), but the same issues; only this time we had to pay a lot more money if we needed the switch re-programmed. As an example, I clearly remember getting a dealer in to set up a new Direct Dial In (DDI) number and Automatic Call Distribution (ACD) group to handle calls for a new advertising campaign. To say it irked us that we had all this technical knowledge in-house, but the very control of our telephone system, the lifeblood of our company, was in the hands of someone else, would be a gross understatement.

What we felt was required was that, ideally, the telephone system would be customer configurable in the same manner as a data networking kit, or at a minimum, have the telephone system installed & maintained by a dealer or reseller, and have the most frequently changed functions customer configurable. Indeed, it appeared to be feasible to take all the various basic voice & data components (hub/switch, Internet access router, remote access server, PBX/Key System), and combine them in a single system. By taking a modular, Lego-like approach, it would allow customers a "one-stop-shop" for all their communication needs along with the ability to simply add LAN or telephone extensions, or enable extra applications (Unified Messaging, Computer Telephony, etc.) as and when required. Utilizing switched digital services, such as ISDN, would allow trunks to be dynamically used for voice and data traffic, requiring less lines then separate voice and data networking equipment, or more efficient use if the same number of lines were to be used. Either way would lead to a reduction in ongoing costs and/or an increase in efficiency.

Having decided on the concept of this "Network Appliance" or "Office-in-a-Box" design, a physical realization was required. A server-based design was a good starting point, after all, we had spent a great deal of time working with Microsoft while we were at Bay Networks and it was very clear that they saw voice and data integration at the desktop as part of their remit. However, while NT offers much for telephony services, the price point associated with the "five nines" reliability associated with PBXs puts server based PBXs into the mid-range price bracket. We saw the market short fall in the entry-level market (SME) and wishing to address this, came up with a design based on datacommunication design values. That is, purpose built hardware, specifically designed to switch voice and data traffic, running custom software; however, all interfaces to the "outside-world" were to be standards based, either de-facto or de-jour. This would allow as much existing equipment as possible to be retained and utilized: analogue telephones, Intel based PCs running Microsoft operating systems, and data networking equipment from other manufacturers. There were to be no fork-lift upgrades with this approach. Putting this together in a 19" rack mountable, stackable module allows for a clean installation, with an easy upgrade path. All user cables for voice and data connect to the front of the units, with all trunk and external connections going via the rear of the unit.

So far we've got the fusion of a traditional PBX with a router, remote access server and LAN hub, and the equipment itself comes in at a lower price point then stand-alone equipment because the duplication of common components, PSUs, cases, trunks, interfaces, etc., can be drastically reduced. Ongoing costs in terms of line rental and administration of adds, moves, changes are reduced and existing equipment can be retained. Good as far as it goes, however, the real "double whammy" with this approach comes from the voice and data integration aspect. It now becomes very easy to take information about an incoming or outgoing phone call and pass it over the LAN to a PC, workstation or server. This is the very basis of Computer Telephony (CT), allowing traditional high-cost extras such as Voice Mail, Unified Messaging, screen popping of databases, screen dialing, etc. to be supplied as standard. It is the delivery of these Customer Relationship Management (CRM) applications to the mass market that was expected to be the clear differentiator for the network appliance. Up until this point CT or CTI, the domain of Call Centers, bespoke systems bringing the voice and data networks together, costing several thousands of dollars per seat.

Network Alchemy was formed on 1st February 1997 with the sole aim of breathing life into the concept of the network appliance. The ArgentOffice was launched in November 1997, with the first customer shipment taking place on 31st January 1998. Since then over 10,000 systems and 135,000 extensions have been installed, while OEM agreements are in place with BT in the UK (BT Fusion) and Iwatsu in Japan (OfficeGate). In between, Network Alchemy was acquired by SDX Business Systems, who in turn were acquired by Lucent Technologies Inc.

While the success of the Network Appliance seems assured for smaller businesses, no one can afford to rest on their laurels. The beauty of the modular approach allows new services, technologies and applications to be embraced, while the customer's original investment is protected. We have launched the ArgentBranch system to cater to larger offices (up to 180 extensions), while utilizing the same extension modules as the ArgentOffice, expanded trunk support to encompass point-to-point digital services and ADSL, added support for Lucent digital phones, for those who (still) don't have or want, a PC on their desk and an S-Bus module to allow stand-alone video conferencing or fax servers to utilise the ISDN trunks. Perhaps more importantly, in terms of future market direction, H.323 gateway support has been added, allowing voice and data networking in the campus (VoIP over the LAN), voice and data networking between sites (VoIP over the WAN) and IP Telephony utilizing

Microsoft's NetMeeting. Before the end of 2000, support for IP hardphones and alternative IP softphone applications will also be available, alongside internal or external H.323 Gatekeeper support.

Lucent Technologies see the Network Appliance fulfilling the roll of the entry-level convergence platform for small to medium sized offices and remote offices of larger organizations. These platforms bridge the gap between yesterday's circuit switched PBXs and Key Systems and the IP only platforms of tomorrow, allowing customers to protect their investment and deliver differentiated IP applications and services to their users as and when required. Analogue, digital and IP soft/hard phones can be mixed on the same system, providing unrivalled investment protection to those who need it most—those with limited budgets and resources.

This Insider Perspective was contributed by Robin Hayman—Alchemy Sales and Marketing Manager, Lucent Technologies. Robin does an incredible job of marketing and evangelizing next generation phone system throughout Europe. As of this writing, the Alchemy Series has shipped over 135,000 extensions, throughout Europe. Thanks, Robin! Copyright (c) Lucent Technologies.

7.2 Praxon Phone Data eXchange

Praxon, Inc. was founded in 1996 and released its Phone Data eXchange (PDX) in 1998. The PDX is designed to be an easy-to-use solution that integrates a variety of communication services and applications—voice, data networking, messaging, and Internet access—into a single turnkey appliance.

Praxon describes the PDX as an Integrated Business Communications (IBC) solution for small to medium-size businesses. Advanced communication solutions have traditionally been beyond the reach of this market segment from both a cost perspective and a technical perspective. In the past, solutions that offered integrated voice communications, data networking,

Figure 7-3: The Praxon PDX Communication Server provides voice communications, data networking, messaging, Internet access, and more in a turnkey appliance. Copyright (c) Praxon, Inc.

and messaging were found only in larger organizations with the budget and technical know-how to support such solutions. Now, with Communication Servers such as the Praxon PDX, even small organizations can take advantage of state-of-art communication capabilities.

The Praxon PDX is currently targeted at businesses with up to 64 internal users. The modular PDX chassis provides three expansion slots, which support from 8 to 64 phone extensions.

7.2.1 Features

What has Praxon packed into their PDX platform? Inside the PDX you'll find a collection of communication applications that is as complete as any offering on the market. The PDX includes:

- A full-featured PBX including support for conferencing, forwarding, Caller ID, DID, music-on-hold, and much more.
- Support for analog or digital (T-1 or ISDN) trunk lines.
- Support for standard analog desktop phones, fax machines, or modems (including automatic inbound fax detection and routing).
- An integrated auto-attendant and voice mail system that allows voice messages to be retrieved by e-mail, Web browser, or telephone. The voice mail system provides support for an unlimited number of "virtual" mailboxes.
- A built-in POP3/SMTP e-mail server that allows you to host your own Internet mail system. E-mail can be retrieved using any standard POP3/SMTP email client.
- A graphical user interface for desktop call control and TAPI 2.1 service provide support to Windows contact managers and other TAPI compliant desktop applications.
- An integrated LAN hub and IP router for Intranet/Internet access.
- A remote access server to provide access to network resources from virtually anywhere.
- Web browser-based setup and administration—plus SNMP management and support for system updates via FTP.
- A modular design that allows you to add features and functions as your business grows.

Figure 7-4: The Virtual Console is a Windows CTI application for the Praxon PDX that provides desktop call control. Copyright (c) Praxon, Inc.

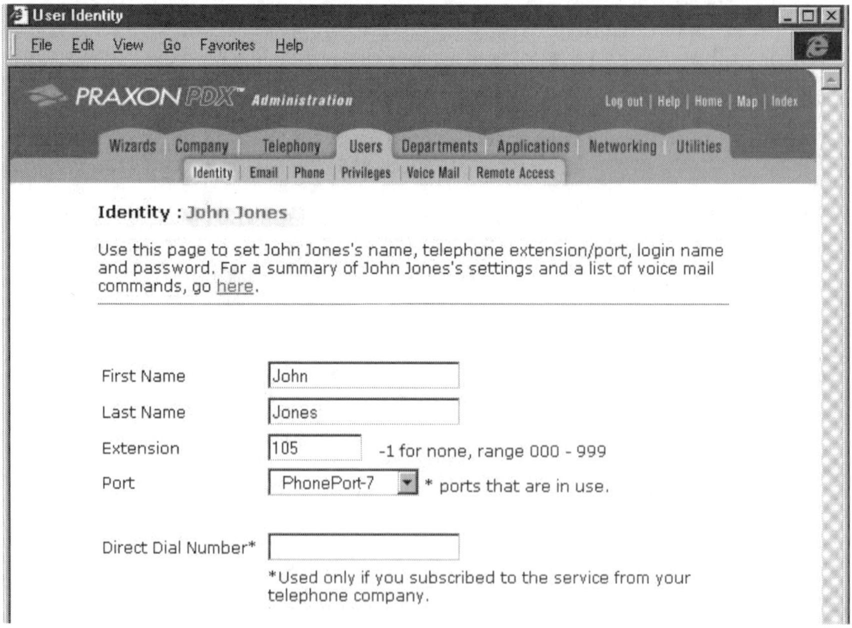

Figure 7-5: The Praxon PDX includes a Web browser-based interface for configuration and administration. Copyright (c) Praxon, Inc.

7.2.2 Summary

Companies looking for a well-rounded solution that integrates all basic communication services into a single turnkey appliance should take a good look at the Praxon PDX.

INSIDER PERSPECTIVE

Integrated Communications for Small and Medium Businesses

By Praxon, Inc.

Advances in technology are driving small and medium businesses toward advanced communication capabilities previously associated with companies many times their size. Although tightened budgets and the increased costs of trained technicians have proved to be obstacles to all but the largest companies, a new path for voice and data communications is emerging that will help establish a level playing field for competitors of all sizes.

The need becomes increasingly clear.

That short deadline purchase order for which your sales representative just flew across the country will not be of much value if locked in a briefcase when fulfillment should be under way. Similarly, documents often need to be drafted and reviewed by numerous people. Clients want the convenience of e-mails and the efficiency of leaving voice mails. Partners want to be able to look up information online. Executives continually search for newer and easier ways to stay "in the loop."

This communications phenomenon has already arrived, and small companies, large corporations, and everyone in-between need to respond. Electronic mail, voice mail, remote access, faxing, paging, and Internet access are quickly becoming basic elements of doing business, much like the telephone itself. Small and medium companies have been forced to navigate the separate offerings of multiple vendors with limited capital and technical expertise; a disadvantage that takes its toll in headaches, anxiety, lost functionality, lost productivity, and frequently a lot of money.

It is to this business segment, "The Forgotten Five Million," that Praxon, Inc. is dedicated.

Opportunity Follows Necessity

A new technology typically begins its life cycle as a luxury, rather than as a necessity. By offering a strong value proposition, larger companies with substantial resources become early adopters, while smaller companies often wait for later versions, lower prices, and a compelling need to counteract their competitors or differentiate themselves.

With modern technology at their disposal, small and medium businesses have an unprecedented opportunity to add new marketplaces, as well as to defend existing ones. Customers established locally, nationally, in other countries, or on other continents are literally on the other end of the wire. These customers seek competitive offerings in hopes of reducing their own costs, but must still screen for other criteria, such as reputability and professionalism. To optimize that position, a company can now present a highly polished image often associated with much larger enterprises. Comprehensive communication options are key.

Integrated Business Communications

The efficiencies of integrated voice and data communications are a compelling business proposition. Computer networks send messages across a wire in the form of digital data. Telephone networks send data across a wire in the form of a human voice. With such strikingly similar missions, it becomes obvious that a company's voice and data communications should be approached as a unified entity.

An industry trend toward computer telephony integration (CTI) has set preliminary groundwork for overall convergence, but fails to meet the full scope of contemporary business needs. In that environment, the desktop computer and the telephone are much more closely interwoven, but users are still left to fend for themselves in messaging, Internet access, and other important features.

These shortcomings in computer telephony integration, combined with the advanced communication needs of today's business, pave the way for a new market segment that addresses the full customer requirement in an integrated fashion. This new segment is called "Integrated Business Communications" (IBC).

IBC addresses four distinct categories: voice telephone service required for both internal and external calls; data networking capability to accommodate file-sharing, peripheral-sharing, and remote access to the office; messaging, which includes e-mail and voice mail; plus secure high-speed Internet access for additional wide area connectivity and access to Web-based resources. Such a feature set should also be supported by simple, browser-based administration that can be managed by support personnel quickly and easily.

The Praxon Phone Data eXchange (PDX)

A technological advancement of the widely-deployed Private Branch eXchange (PBX), the Praxon PDX establishes a new paradigm in the communications marketplace, bringing voice and data together. The PDX is a scalable, integrated communications platform for small and medium companies, plus the remote offices of larger organizations. This easy-to-use, affordable product offers important features and services today, and can be adapted to future technologies with the addition of new modules.

The Praxon Phone Data eXchange combines PBX features, such as call management, caller ID, and international ringing cadences, with a full complement of data networking, messaging, Internet, and administrative functions onto an integrated platform. The solution meets the needs of professional offices, small manufacturers, research-and-development organizations, and similar entities.

Features include an automated telephone attendant, peripheral- and document-sharing, voice mail, e-mail, remote access, Internet access, and data backup. In addition, the Praxon PDX serves as a multimedia mailbox for playback, storage, and forwarding of voice mail and e-mail messages. This capability eliminates the expense of buying separate products to accommodate each feature.

A Simple, Effective Solution

Conventional computer networking has added a new layer of complexity to the elegant, pick-up-and-dial voice solution. Users, or their support staff, have had to grapple with various operating systems and diverse protocols, as well as evaluate a handful of options for meeting the same objective, weighing the pros and cons of the individual technology. As is the case with the voice telephone, the person hitting the "send" button in a data transmission involving the Praxon Phone Data eXchange does not need to understand the complexities of moving data from point to point. The Praxon PDX is designed to be seamless and transparent to the end user. Designed with ease of use as a top priority, the Praxon PDX features a point-and-click interface that guides the user through the system straightforwardly and efficiently. In addition, future adds, moves, and changes to the personnel roster can be completed within one minute, without the assistance of a specially trained support person.

The Praxon PDX is compatible with all popular operating systems, including Windows 95, Windows 98, Windows NT, Unix, and Mac OS. The Praxon PDX does not require an external operating system. Its own OS, which was developed specifically for both voice and data, is embedded in the product, itself. This robust software platform fully supports the features offered by the Praxon Phone Data eXchange.

Although the precise level of financial savings can vary from site to site based on individual circumstances and selected options, the Praxon PDX typically pays for itself within 12 months. The Praxon PDX enables peripherals to be shared, reduces the number of necessary phone lines, and provides a unified location where e-mails and voice mail messages can be saved for each individual user.

The Praxon Phone Data eXchange in Action

Setting up your business communications (i.e. telephones, computers, etc.) can be a difficult, costly, and frustrating experience—especially for small and medium size businesses lacking extensive in-house technical resources.

At most offices, each desk has its own telephone, which is typically connected to a key system or, Private Branch eXchange (PBX) serving as a central "switchboard" for the entire location. An additional voice mail system might be necessary, along with an automated attendant.

Sitting next to the telephones, most desks also have computers, which are connected to one another using hubs, bridges, routers, and switches. Peripherals, such as printers, are connected to the local area network using the same types of devices. Remote access and Internet connections typically require a modem or access server. As modern business communication requirements continued to increase, the need to purchase special servers supporting voice mail and email emerged as well.

Selecting the right solution that offered the full range of communication features in a straightforward, comprehensive, and cost-effective manner became a major challenge for executives- especially those within companies that did not have technical expertise beyond their core competencies. Making all of the separate products work together was a complicated task. Buying each device separately made it virtually impossible to realize economies of scale. To make matters worse, it was often difficult to pinpoint the exact source of malfunctions, leading to substantial finger pointing when something went wrong.

The Praxon Phone Data eXchange has changed all that.

The Praxon PDX is an all-inclusive Integrated Business Communications device that combines voice, data networking, messaging, and Internet access into an affordable product for small and medium businesses.

Rather than buying an assortment of equipment that is probably not specifically designed to work together, the Praxon Phone Data eXchange is a comprehensive platform connecting your standard office equipment to integrated voice mail, e-mail, secure Internet access, remote access, file sharing, and document sharing capabilities. The Praxon Phone Data eXchange is compatible with major operating systems, (including NT, Windows 95, Windows 98, Unix, MacOS), standard analog phones, personal computers, and leading e-mail programs.

The Praxon Phone Data eXchange enables voice communications and computer communications to share the same outside lines, eliminating the need to pay for separate dedicated ones. This capability can enable users to reduce their number of outside lines and save money, typically recouping the total PDX investment of $300 to $425 per seat in about 12 months. Capital investment can also be reduced in startup situations because the customer no longer needs to bear the expense of separate, stand-alone networking components, such as hubs and routers.

Designed with ease-of-use as a top priority, initial configuration of the Praxon Phone Data eXchange takes only 10 minutes, and subsequent adds/moves/changes can be input within one minute from a simple graphical user interface without a specially trained support person.

Developed especially for small and medium size businesses, the Praxon Phone Data eXchange significantly lowers operating costs while providing integrated, full-featured communications capabilities in an easy-to-use, straightforward manner.

This Insider Perspective was contributed by Praxon, Inc. Many thanks to Joyce Tam and the folks at Praxon for permission to publish this article. Copyright (c) Praxon, Inc.

7.3 Vertical Networks InstantOffice

Vertical Networks, Inc. shipped its first system in October 1999 with the goal of offering small and medium size offices the same leading-edge voice and data communication capabilities found only in large corporations. To provide a comprehensive, easy-to-use, and cost-effective solution, Vertical Networks consolidated a variety of communication services into a single integrated platform. According to Vertical Networks, its InstantOffice product line is the first of a new class of communication solutions known as Integrated Communications Platforms (ICP).

> "An Integrated Communications Platform (ICP) is a new class of communications solution that combines voice functionality (circuit- and packet-switched PBX, VoIP gateway, voice mail, voice mail networking, and automated attendant) and data functionality (multiprotocol router, Ethernet LAN ports, remote access server, and multiservice WAN modules) as well as support for third-party applications. The platform delivers truly integrated voice and data networking that is easier to manage and has a much lower total cost of ownership than traditional communications infrastructures, making it the ideal platform for branch offices and small businesses. An ICP not only tightly integrates many features and functions into one system but also includes flexible and cost-effective remote diagnostic and management tools."

One of the unique features of Vertical Networks InstantOffice is the control system that it uses to achieve this integration and support for third-party solutions. While most vendors of Communication Servers have implemented a proprietary real-time control system or an embedded version of UNIX, Vertical Networks is running Microsoft Windows NT on an Intel Pentium platform. Although InstantOffice runs Windows NT, the hardware platform does not conform to the standard PC reference architecture. Rather, InstantOffice is a purpose-built hardware platform, designed specifically to provide reliable communication services.

By using an embedded version of Windows NT, Vertical Networks has leveraged a mature, reliable, and readily-available operating system. Additionally, Windows NT offers a feature-rich development platform that is familiar to a large pool of talented developers. Windows NT also allows Vertical Networks to take advantage of third party applications designed for this platform, such as Xiox Corp.'s General Business Software call-accounting package.

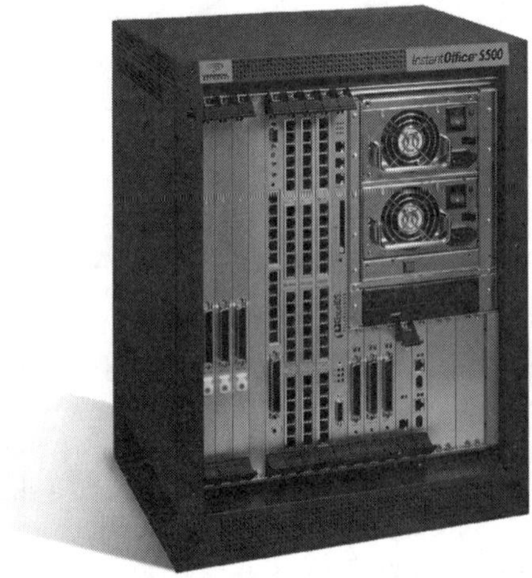

Figure 7-6: Vertical Networks InstantOffice is based on a purpose-built hardware platform and runs an embedded version of Microsoft Windows NT. Copyright (c) Vertical Networks, Inc.

Vertical Networks InstantOffice scales up to 84 phone ports and 84 data ports in a single chassis, making it suitable for most small to medium size businesses and branch offices. Additionally, multiple InstantOffice units can be linked via an embedded multiprotocol IP telephony gateway, which can be used to provide a "virtual-tie-line" between multiple locations.

7.3.1 Features

Vertical Networks InstantOffice delivers an impressive collection of voice and data communication features in a single platform. These features include:

- Circuit-switched and packet-switched PBX—InstantOffice supports full PBX switching services on traditional circuit-switched voice. With an embedded voice over IP (VoIP) gateway, the PBX functionality can be seamlessly applied to packet-switched networks as well.
- Auto-attendant voice mail and AMIS voice mail networking—InstantOffice includes an auto-attendant and voice mail system. The voice mail system includes the ability to forward voice messages as WAV attachments to e-mail messages and also supports voice mail networking based on the Audio Messaging Information Specification (AMIS).
- Computer telephony applications—If you're not using one of the multi-line digital feature phones supported by Vertical Networks, you can use a PC-based application, OfficeCommunicator, for real-time call control and access to voice mail messages. InstantOffice also supports TAPI, giving you the ability to run third-party desktop computer telephony applications.
- Internet/intranet access—InstantOffice provides a 10/100 LAN hub and supports standard network protocols, allowing you to deploy a high-performance LAN. With WAN support for T1, ISDN PRI, frame relay, SDSL, 56/64K DDS, and ana-

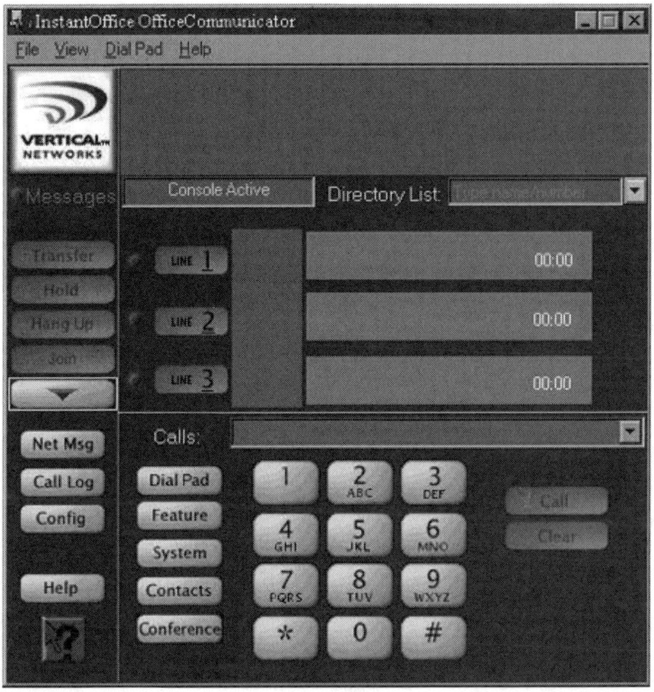

Figure 7-7: A desktop computer telephony application provides users with graphical call control. Copyright (c) Vertical Networks, Inc.

Figure 7-8: A browser-based management console allows InstantOffice to be administered and configured from virtually anywhere—especially useful for branch office installations. Copyright (c) Vertical Networks, Inc.

log lines, the InstantOffice platform can be networked with other sites or the Internet. The WAN support is optimized to conserve network bandwidth using a variety of techniques including; dial-on-demand routing, compression, filtering, and spoofing.

- Security and administration services—These services include a built-in firewall, encryption, DHCP, DNS, and NAT services. InstantOffice also supports SMTP management, customizable access levels, and a Web-based interface for remote access and troubleshooting. The Web-based administration interface is a critical feature that allows headquarters-based administrators to support and configure branch office installations.

For enhanced reliability, InstantOffice optionally includes redundant power supplies and/or hard drives (RAID-1). The system comes in rack-mount or stand-alone configurations.

Digital Feature Phones

In addition to standard analog desktop telephones, Vertical Networks also supports their own digital feature phones. For many phone system users, the telephone *is* the phone system. The features available from the desktop telephone are what many users care about most. A multi-line display screen, message waiting lights, the ability to monitor and answer calls on multiple extensions, one-button access to advanced features, etc. are commonplace on traditional PBX telephone sets. When implementing a next generation solution, many users will demand the sophisticated features that were available to them on a traditional PBX telephone.

Not only Vertical Networks, but almost all next generation vendors, recognize the importance of providing advanced feature phones. Through their own development efforts or through partnerships with third-party vendors, next generation phone system manufacturers are beginning to support desktop feature phones.

Vertical Networks offers 4 different models of digital feature phones with up to 64 programmable feature keys, a 2-line x 24-character display panel, a built-in speakerphone, and more. The Vertical Networks phones allow a single extension to appear on multiple phones and for multiple extensions to appear on a single telephone. With Direct Station Select (DSS) and the Busy Lamp Field (BLF) features, Vertical Networks telephones can also be used by attendants to efficiently handle and dispatch incoming calls.

7.3.2 Summary

Not only is InstantOffice an integrated communications platform, but it is also a modular platform designed to support organizations that are migrating to a converged infrastructure. Accordingly, if your small business requires just a basic phone system today, you can purchase only the PBX components of InstantOffice. Then, when you need the data networking support, you can easily enhance your InstantOffice platform with those features.

InstantOffice is packaged in two different configurations, the 3000 series, which supports from 5 to 36 users, and the 5000 series, which supports from 36 to 84 users. Systems are sold through VARS, system integrators, and carriers. Small companies and branch offices looking for an easy-to-use, integrated communications platform will find InstantOffice an attractive option.

INSIDER PERSPECTIVE

The Road to Voice/Data Convergence

By Vertical Networks

Computers, networks, and the Internet have brought vast improvements in business and personal productivity, but until recently, voice and data have required separate office networks. Voice traffic has traveled over voice networks, and data has traveled over packet-based networks. Advances in IP and carrier network technology, however, are changing this picture. New IP telephony technology has made it possible to transport voice traffic over data networks, and new carrier technology is enabling converged broadband voice/data services. Today, the world is moving quickly toward a convergence of voice and data networks. We can see a future when all networks will be based on broadband digital technology and packet-based protocols, bringing to century-old voice networks the same surge of new applications and productivity that it has to computers and other digital solutions.

Network convergence will ultimately bring far simpler and more manageable voice and data systems for carriers, Internet Service Providers (ISPs), and customers alike. Since the telecommunications business was deregulated in the mid-1980s, we have seen an explosion of new companies, new technologies, and new product solutions. As a result, everyone seeking to buy and use converged network solutions faces more and more product choices and increasing network complexity.

This paper offers a roadmap to voice/data convergence for small companies and enterprise branch offices. It describes why voice/data network convergence is happening, what it means, and how small companies and corporate branch offices can plan for it.

Why Convergence is Inevitable

The Public Switched Telephone Network (PSTN) is an analog system that was designed to carry voice calls. Early data communications was done primarily using the PSTN via dial-up calls and modems, which converted digital traffic into analog traffic (or vice versa) for transport over the PSTN. As data communications became increasingly important to businesses of all sizes in the 1980s and 1990s, telecommunication carriers discovered that they needed more network capacity to handle the additional data traffic. Dial-up data calls for Internet or e-mail access could easily last 30 minutes-far longer than most voice calls-and a PSTN designed to handle voice-only traffic was clearly inadequate.

To meet carrier needs for more capacity, companies like Ascend Communications, Cascade, and Stratacom emerged with solutions that allowed carriers to create separate "overlay" public networks for data traffic in order to increase their capacity. Carriers eagerly added these data-handling capabilities to their networks. Access technologies such as T1 service, ISDN (Integrated Services Digital Network), and DSL (Digital Subscriber Line) have given small business customers faster and more reliable data connections.

Overlay networks were a logical solution for carriers with billions invested in their analog PSTN plants, as the overlays did not disrupt the PSTN's voice-carrying functions. But by creating an essentially separate data network, carriers were also creating higher construction, maintenance, and provisioning costs as well as more network complexity. At the same time, business customers with growing data-handling needs were also forced to adopt separate voice and data solutions- PBX systems and voice lines for voice traffic, voice mail, and automated attendant features, for example, and data lines, IP routers, hubs, local area network (LAN) wiring for data service within the office, and wide area network (WAN) connections for access to the outside world.

The end result has been better data services, but at the cost of much greater complexity and higher costs for carriers and their customers. Typically, each new voice or data component means adding a new device to the network, so the complexity and cost problems continue to grow. The obvious solution is to converge voice and data services on one network. Recent advances in IP voice technology have now made it possible to carry all voice and data services over one IP-based network, and carriers and their customers are moving in that direction.

The Benefits of Convergence

True convergence of voice and data services on one, IP-based network will bring a number of immediate and longer-term benefits to carriers and their customers. Putting voice and data on the same network is a win for everybody.

Lower equipment costs

For carriers, the major cost of broadband service is in the physical copper assets at the central office facility. It makes sense to leverage the physical copper pair by bundling as many services as possible. In order to maintain today's separate voice and data networks, carriers must invest in gateways, digital access concentrators, and other pieces of equipment to merge data traffic with the PSTN. Frequently, each of these boxes requires its own management system, and carriers must develop expertise in it. Carriers' networks are now so complex that no single technical staff member understands the whole picture. By handling all traffic on a single network, carriers can greatly simplify their networks and eliminate whole layers of equipment and the expense of managing them.

Separate data and voice facilities also means higher costs for business customers. A company might have a PBX and separate voice mail for voice services along with routers, switches, hubs, channel bank, firewall, and virtual private network (VPN) equipment to handle data traffic. As with carriers, customers must often deal with a different management system for each box, and they must hire or train technical staff to support each new component. In addition, all those boxes occupy space, which is often at a premium in smaller offices. With a single, converged network that runs voice and data, customers can provide, manage, and support both sets of services on a single infrastructure. By reducing the number of support components required, businesses can also reduce space requirements.

Lower Service Costs

Converged networks mean less expensive voice and data communication services from providers. With all traffic flowing over one network, customers can eliminate multiple service connections and wasted capacity. Today, for example, an office might have a T1 connection for data from one provider along with 12 incoming analog lines for telephone service (from another provider) to implement a main number, a fax machine, and direct inward dialing (DID) service for five employees.

In this scenario, the customer is paying for a dozen individual voice lines that are underutilized, especially at night, when that capacity might be used for data services (overnight backups, for example). If the customer needs more data bandwidth, it must pay for an additional data line, even though it has excess capacity for its voice lines.

A converged network could deliver all services via one connection to one provider. Software management tools can then be used to provision data and voice services such as DID lines very flexibly. In many cases, companies can be more productive with fewer outside service connections because they are using the available capacity more wisely.

Lower Maintenance and Operations Costs

Converged networks are far simpler, requiring fewer pieces of equipment and fewer management systems. By simplifying overall network design and operations, converged networks lower technical and management personnel costs. Customers additionally lower their administrative costs: with only one provider, they have only one bill to process and one provider relationship to maintain.

New Services

Customers want access to new voice and data services such as VPNs, voice-over-IP (VoIP) for carrier services, VoIP for branch office "virtual" tie lines, and voice mail networking. Carriers and providers want to rapidly deploy these new, high-value services.

With a converged network, developers can work with a single, common set of open standards to create the new services customers want. We have seen the explosion in applications and services fostered by Internet standards, and we can expect the same proliferation in converged voice and data networks. In addition, customers who obtain all their voice and data services through one broadband connection will easily be able to access new services as they become available-again, without additional costs for equipment.

Requirements for Convergence

Carriers have already begun migrating to converged voice and data networks. Since the largest cost of deploying service is the physical copper infrastructure needed to reach customers, carriers are now finding ways to merge voice services into data service networks in order to reduce provisioning expenses. For example, carriers are rolling out converged access services like T1 and DSL, which can be provisioned for voice or data.

But delivering voice and data over one line to the customer is only the beginning of convergence. If customer premises equipment (CPE) is still configured for separate voice and data services, the customer doesn't benefit from the carrier's newly converged offerings. For example, ISDN has been around for a decade or so, but it has never reached its expected penetration partly because there haven't been any cost-effective CPE platforms that allow customers to fully leverage an integrated voice/data connection. In order to leverage convergence without an integrated CPE platform in the past, customers had to buy a channel bank and manually split out DS0 lines, adding more complexity to their office telecommunications infrastructures.

Understandably, customers saw little value in convergence, and the carriers offering converged services couldn't easily profit from them. ISDN and T1 connections were consequently used for voice, data, or video services, but not all three.

ICPs: The Convergence Solution for Small Offices

To truly leverage converged voice and data services from carriers, customers need a new type of communications solution, the integrated communications platform, or ICP. An ICP delivers all of the benefits of convergence by providing all voice and data services in a single, compact platform. An ICP should be modular so companies can buy only the equipment they need, and it should expand easily to support additional users or features in the future.

Vertical Networks pioneered the ICP with the InstantOffice(tm) system, the world's first integrated voice/data CPE platform, in 1998. Today, Vertical is the leading ICP supplier, and it is the only provider of ICPs with a corporate focus on full convergence.

The Customer Convergence Roadmap

Vertical's vision for convergence is one that enables small and remote offices to simplify their communications infrastructures as their needs allow, migrating from separate carrier voice and data connections to a fully converged, end-to-end IP network that extends from the desktop all the way to the telecommunications network core. The migration to fully converged, IP-based voice/data networks is occurring in five phases, as follows:

1. Platform and Management convergence
2. Access convergence
3. Service convergence
4. Applications convergence
5. End-to-end convergence

With the InstantOffice system, customers can easily set out on a migration path that improves management, lowers costs, and increases network efficiency in every phase. This section details these phases and explains how small companies as well as enterprise branch offices can use ICPs like the InstantOffice system to make the most of each phase.

Phase 1: Platform and Management Convergence

The first phase of convergence is to physically merge separate voice and data services from carriers in a single CPE platform, the ICP. Nearly all customers today still maintain separate WAN and voice (PSTN) services from carriers, but the ICP allows them to integrate these services at the premises.

The ICP replaces multiple data and telephony platforms with a single, integrated solution. Rather than having to install, configure, and maintain separate routers, hubs, PBXs, and other solutions, customers can use the ICP to handle all of these functions in one compact, modular device.

The InstantOffice ICP connects to both PSTN and data networks, providing multi-line voice service along with high-bandwidth Internet/intranet access. It offers all of the key voice and data services that offices need, including a PBX, automated attendant, voice mail system, VoIP gateway, 10/100 switched Ethernet hub, remote access server, and multi-service WAN connectivity modules in one platform. Customers benefit from the reduced costs of converged services and reduced management.

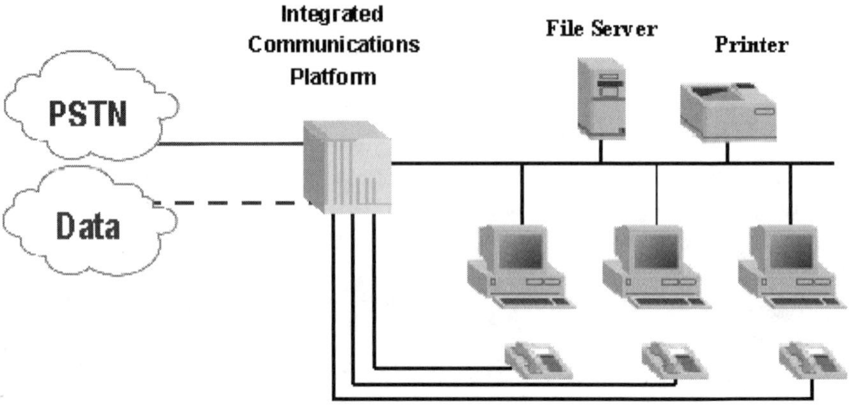

Phase 2: Access Convergence

In Phase 2, convergence extends to the physical connection that provides access from the carrier. Rather than delivering voice and data services on separate connections, carriers use a single connection to bring them to the customer.

In this phase, carriers offer integrated, broadband voice and data services such as T1, ISDN, and DSL. Lower carrier costs to deliver integrated voice and data services on a single copper connection are passed on to customers, who also benefit from lower administrative costs because they have just one voice/data provider rather than two. There is a substantial cost savings when using an ICP in Phases 1 and 2 compared with a traditional infrastructure that uses separate voice and data solutions. The ICP supports Phase 2 with access modules that handle converged broadband access services from carriers.

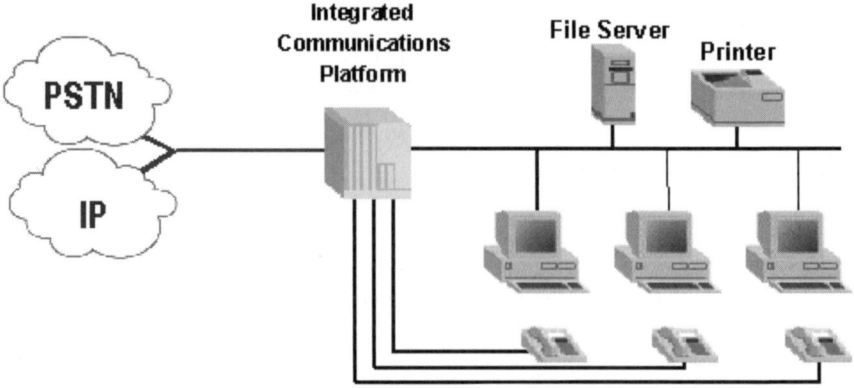

Phase 3: Service Convergence

With physical convergence achieved from the carrier network through to the ICP, it becomes possible to deliver both voice and data services using the IP protocol. Using a single, universally accepted protocol allows customers and carriers to obtain more value from the converged connection beyond just physical access. All next-generation carriers are moving to support IP as a single communications protocol for voice and data. In Phase 3, telephone connections are still delivered from the ICP on separate physical networks within the office, but customers will be able to migrate to new IP phones that offer simplified management and more features. (If they prefer, customers can continue to leverage their existing investments in analog and digital telephones.)

By converging all voice and data service on the IP protocol, customers can simplify and reduce the cost of existing services and enable new services that would have been impossible before, like the following examples:

- DID numbers that once required separate incoming and outgoing analog lines for each user can be implemented on a single line on a converged IP network
- Customers can layer VoIP on a T1 or DSL connection, creating virtual tie lines between remote offices and the main office that save money by bypassing the PSTN toll carrier.
- Voice mail can easily be networked among other remote offices, bringing branch office voice mail into the corporate environment and allowing mobile employees to obtain messages from any office without paying toll charges or using leased voice lines.

Customers also benefit from further reductions in operational overhead. With one ICP platform, all voice and data functions such as user setup, performance monitoring, and data network configuration can be handled through a single management interface.

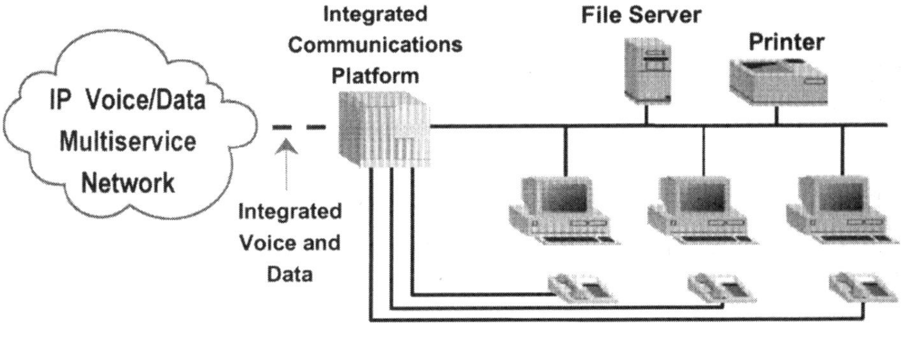

Phase 4: Applications Convergence

In Phase 4, we begin converging once-separate voice and telephony applications. In recent years, cellular telephones, pagers, wireless PDAs, and handheld computers have offered the promise of mobile productivity, and Phase 4 allows customers to realize that promise.

In the beginning, customers will want to focus on "workhorse" applications such as voice mail and e-mail. With voice/data convergence, customers will allow their employees to download voice mail to an IP phone, handheld computer, or wireless PDA.

As Phase 4 continues, dozens of developers will take advantage of the convergence on IP standards to create new applications that will further improve productivity by supporting voice, data, or video over a common communications infrastructure. We have so far seen only glimpses of these New World applications, but they will eventually transform all areas of business.

Whatever develops in the future, however, the ICP is the point of convergence, the glue that provides access to services and applications, whether hosted at the small office, at the branch office, at the enterprise headquarters, or on the Internet.

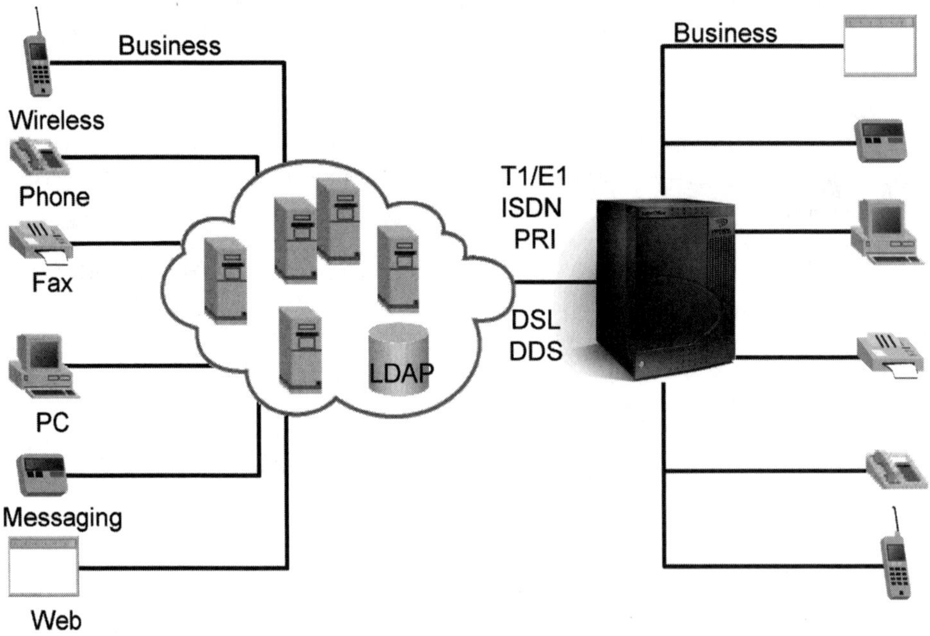

Phase 5: End-to-End Convergence

This phase will complete the migration to converged voice/data networks. All voice and data traffic inside the customer premises will be carried on a single IP-based LAN, effectively ending the transitional era of computer-telephony integration. (When all voice and data traffic is carried on IP networks, there is no longer any barrier between computers and telephones.) Every device-including telephones, PCs, printers, file servers, fax machines, PDAs, pagers, and cellular phones-will have an IP address.

The converged LAN will be a single, fully manageable platform for all devices, all services, and all applications. Desktops will link to the voice/telephony network with a single connection that supports PCs as well as IP telephones. It will be possible to expand telephone service without adding new line cards to the ICP, as would be necessary with a PBX system. Remote users will be able to access and control their desktop PCs via IP cell phones.

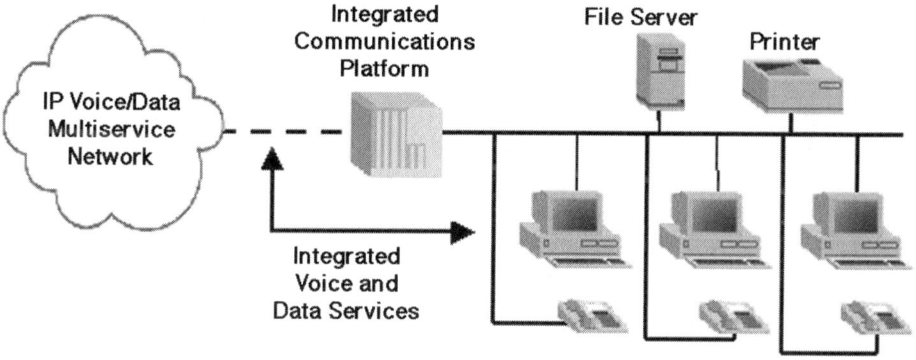

With all devices on a single IP network, it will be far easier to manage user connections via a single, Web-based management interface. From adding or deleting users to monitoring network performance at each desktop, a converged IP network will offer complete flexibility for service deployment at a lower cost than ever before. In addition, full convergence will spawn even faster development of new applications such as enterprise-based softswitches for bandwidth control and network traffic optimization.

Vertical Networks: Convergence with Confidence

Completely converged voice/data networks based on IP are coming, and customers need CPE solutions that support smooth and cost-effective migration at each phase. ICPs are the customer solution for convergence, and for maximum reliability and flexibility, customers should choose an ICP vendor that supports every phase of convergence with robust, modular, and scalable solutions.

Vertical Networks is the leading supplier of ICPs, and is the only ICP vendor whose primary corporate mission is to enable full voice/data convergence for small businesses and enterprise branch offices. Vertical Networks entered the ICP category with its debut of the InstantOffice system in 1998. It was the first to integrate a VoIP gateway with a circuit-switched PBX and to support integrated broadband access with InstantOffice 2.0 in 1999. In 2000, Vertical Networks announced the expanded

InstantOffice family of ICPs, which supports third-party applications-such as firewalls, VPNs, and call centers-and supports both analog and digital phones.

The InstantOffice system is a modular voice, data, and applications platform that combines the functionality of multiple communications products into a tightly integrated, remotely manageable solution, including a PBX, voice mail system, VoIP gateway, multiprotocol router, 10/100 Ethernet LAN ports, remote access server, and multiservice WAN modules.

With a large and proven installed base, top-tier investors and partners, and strong carrier relationships, Vertical Networks is the most advanced, most reliable provider of ICPs for small and remote offices. Vertical Networks offers legacy integration and cost savings at each phase of the migration process while building a system for the future, allowing customers to make a smooth transition to full convergence at their own pace.

This Insider Perspective was contributed by Vertical Networks. Thanks to the folks at Vertical Networks for permission to publish this Convergence Roadmap. Copyright (c) Vertical Networks, Inc.

Chapter 8

The IP-PBX

8.1 Introduction

Of the different classes of phone systems discussed in this book, the IP-PBX represents the most significant departure from the traditional PBX architecture—and, arguably, it is the one that offers the most opportunity for the future.

With a traditional PBX (and even with many next generation solutions), the major phone system components—the switching matrix and the call control system—are tightly coupled. Usually these components are contained within a single cabinet, linked by a proprietary bus. An IP-PBX, on the other hand, uses the switching fabric of an IP data network. The system components, such as the control system or media processing applications, are distributed throughout the network on either PC platforms or within network appliances. The various components of the system send call control information and media over the IP network using standard and emerging protocols. Integration with legacy circuit-switched networks is provided by special-purpose media gateways, also connected to the IP network.

The distributed architecture of the IP-PBX, along with its modularity and reliance on networking and computing technologies, makes this class of solutions the most radical departure from the traditional PBX design.

8.2 Market Trends

IP-PBXs are rewriting the rules of business voice communication solutions. The traditional PBX was simply an extension of the same technology used by the phone company's central office and throughout the PSTN. Rather then looking to the PSTN for inspiration and direction, the IP-PBX vendors have taken their cue from the Internet and the innovation and evolution of data networks. The trends currently shaping the evolution of the IP-PBX industry are discussed below.

- Standards are a major focus for the industry. There's universal recognition that the benefits of a distributed, modular architecture will remain compromised until standards for interoperability are agreed upon and widely implemented.

  With true interoperability in place, customers will be able to mix and match system components and advanced applications to meet their needs. Standards-based interoperability allows customers to create true multi-vendor solutions, rather than relying on the proprietary offerings of a single vendor.

- The virtual IP-PBX is emerging. With the distributed architecture of an IP-PBX, the call control system can be located almost anywhere on an IP network. A number of companies, such as Netergy Networks (see section 11.2), have taken advantage of this architecture and are offering solutions that allows service providers to host IP-PBXs and offer phone system service to subscribers.

- The features of IP-PBX desktop phones are second to none. In Chapter 2 we mentioned that one of the shortcomings of next generation solutions is that the desktop telephones cannot match the sophistication and features offered by traditional PBX manufacturers. However, there is one exception: the IP-PBX phone. The IP-PBX is inherently a distributed solution—typically there is no centralized server that concentrates system features and functionality. Accordingly, IP-PBX manufacturers have been primarily concerned with making the telephone a sophisticated, intelligent network device. Some third-party vendors of IP telephones, such as Pingtel, have gone so far as to add support for Java applets in their phones.

- IP-PBXs are typically more scalable than other next generation voice communication solutions. Solutions such as Cisco's Enterprise Telephony System are designed to support thousands of users and compete with traditional PBXs for the enterprise market.

8.3 Relevant Standards

The standards which are most important to the IP-PBX marketplace are the same standards driving the Communication Server marketplace and, to a certain degree, the PC-PBX marketplace. The standards driving this marketplace are designed to provide interoperability between voice communication devices distributed throughout a packet-switched network environment. These standards include ITU H.323 (see Section 4.3), IETF SIP (see Section 6.3), and IETF MGCP.

Standards Tutorial: Media Gateway Control Protocol (MGCP) Overview

By Danny Levin—Vice President of Engineering, RADVision, Inc.

MGCP-based Voice over IP solutions separate call control (signaling) intelligence and media handling. In an IP telephony network that spans circuit-switched networks and IP networks, an MGCP-compliant Media Gateway Controller is responsible for controlling the resources of separate MGCP-compliant Media Gateways. The Media Gateway Controller also handles call processing by interfacing with the circuit-switch network via an optional signaling gateway and the IP network via communications with an IP signaling device such as an H.323 gatekeeper. With this architecture, it is possible to build extremely cost-effective Voice over IP Media Gateways.

Various bodies are working closely on expanding Media Gateway Control architectures. As a participant in this joint effort, RADVision is able to design its MGCP toolkits in such a way as to allow for easy migration to comply with the emerging H.248/MEGACO protocol, which is being jointly defined by the ITU and IETF. RADVision's first generation MGCP toolkits are based on the IETF MGCP specification and support the MGCP/NCS profile adopted by the CableLabsTM consortium.

Brief Introduction to MGCP

MGCP is used for the control of telephony gateways from external call control elements called Media Gateway Controllers. A telephony gateway is a network element that performs conversion between the audio signals carried on telephone circuits and data packets carried over the Internet or over other packet networks.

Examples of gateways are:

- Trunking gateways that interface the telephone network and VoIP network.
- Residential gateways that provide traditional analog (RJ11) interfaces to Voice over IP networks.
- Access gateways that provide traditional analog (RJ11) or digital PBX interfaces to Voice over IP networks.

Evolution of Media Gateways

Despite the widespread support of H.323 by vendors, the protocols used for call control within IP networks and between IP-based networks and the PSTN are currently being studied by a number of organizations. Towards the end of November 1998, Bellcore and Cisco announced that their proposed Simple Gateway Control Protocol (SGCP) and the Internet Protocol Device Control (IPDC) protocol suite had merged to become the Media Gateway Control Protocol (MGCP). The MGCP draft has been submitted to the IETF and the European Telecommunications Standards Institute (ETSI) Project TIPHON (Telecommunications and Internet Protocol Harmonization Over Networks) for consideration. Both the IETF and ETSI are working toward defining end-to-end interoperability between IP-based networks and the PSTN.

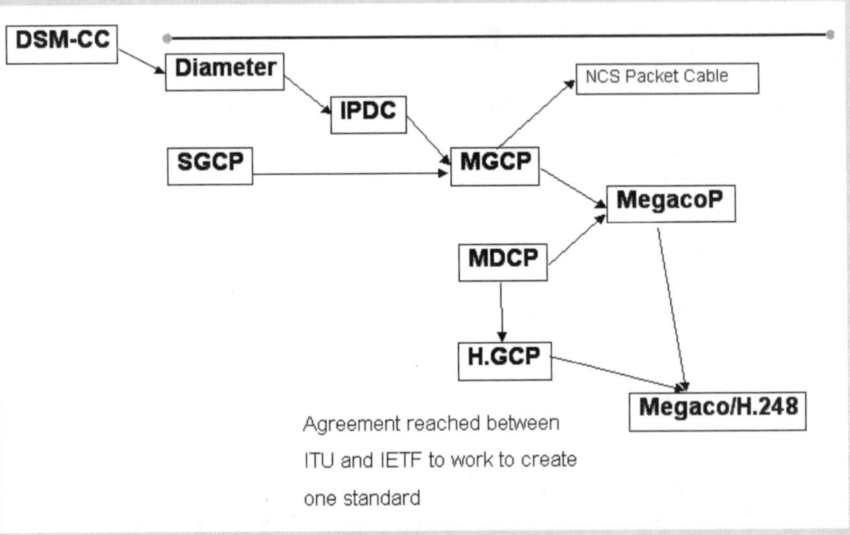

The Evolution of MGCP

Another player, CableLabs, is leading the PacketCable project to define standards for IP-based voice over cable.

After the divergence of MegacoP and H.GCP an agreement between the MEGACO group in IETF and the SG-16 group in ITU-T has been reached to cooperate and the result is a a single protocol, MEGACO/H.248.

The MGCP Model

MGCP is designed as an internal protocol within a distributed system that appears to the outside as a single VoIP gateway. This system is composed of a Media Gateway Controller, that may or may not be distributed over several computer platforms, and of a set of gateways, including at least one Media Gateway, that perform the conversion of media signals between circuits and packets. In a typical configuration, this distributed gateway system will interface on one side with one or more telephony switches (circuit switches), and on the other side with an H.323 (or SIP) conformant system.

In the MGCP model, the gateways focus on the audio signal translation function, while the Media Gateway Controller handles the signaling and call processing functions. Using an H.323 analogy, the Media Gateway Controller implements the "signaling" layers of H.323 and presents itself as an "H.323 Gatekeeper" or as one or more "H.323 Endpoints".

Architecture

MGCP describes a call control architecture where intelligence of the call control is outside the gateways and handled by external call control elements (Call Agents). The MGCP assumes that these call control elements will synchronize with each other sending coherent commands to the gateways under their control. MGCP does not define a mechanism for synchronizing Call Agents. MGCP is a master/slave protocol where the gateways are expected to execute commands sent by the Call Agents.

MGCP main entities are:

- Endpoints
- Connections

MGCP assumes a connection model where the basic constructs are endpoints and connections. Endpoints are sources or sinks of data and can be physical or virtual. Physical endpoint creation requires hardware installation while software is sufficient for creating a virtual endpoint. An interface on a

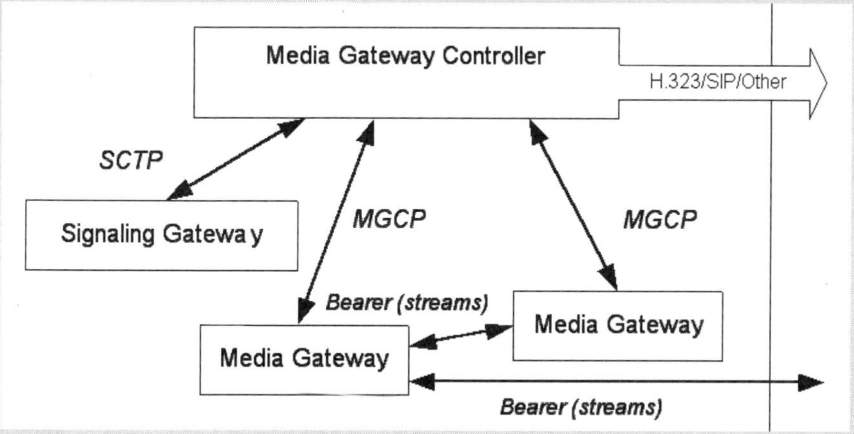

gateway that terminates a trunk connected to a PSTN switch is an example of a physical endpoint. An audio source in an audio-content server is an example of a virtual endpoint.

MGCP Architecture

Connections may be either point-to-point or multipoint. A point to point connection associates two endpoints. Once this association is established for both endpoints, data transfer between these endpoints can begin. A multipoint connection is established by connecting the endpoint to a multipoint session.

Connections can be established over several types of bearer networks: audio packet transmission using RTP and UDP over a TCP/IP network, audio packet transmission using AAL2 or another adaptation layer over an ATM network, and transmission of packets over an internal connection. For both point-to-point and multipoint connections the endpoints can be in separate gateways or in the same gateway.

The MGCP commands include:

- Endpoint Configuration
- Notification Request
- Notify
- Create Connection
- Modify Connection
- Delete Connection
- Audit Endpoint
- Audit Connection
- Restart Progress
- Move Connection

H.323 versus MGCP

In an MGCP architecture there is a tight coupling between the Media Gateway (endpoint) and the Media Gateway Controller (server) in a Master/Slave relationship. The server is mandatory and manages the calls and conferences and supports the services provided. Control is via signals and events. The Media Gateway is unaware of the calls and conferences and does not maintain call states.

In an H.323 architecture there is a loose coupling between the endpoint and the server (gatekeeper). The server is optional and the system is updated dynamically (in real-time). The endpoint is powerful and intelligent and it terminates signaling, control and bearer streams. The endpoint is aware of calls and conferences and the services provided. It maintains the call states but does not expose its internal structure.

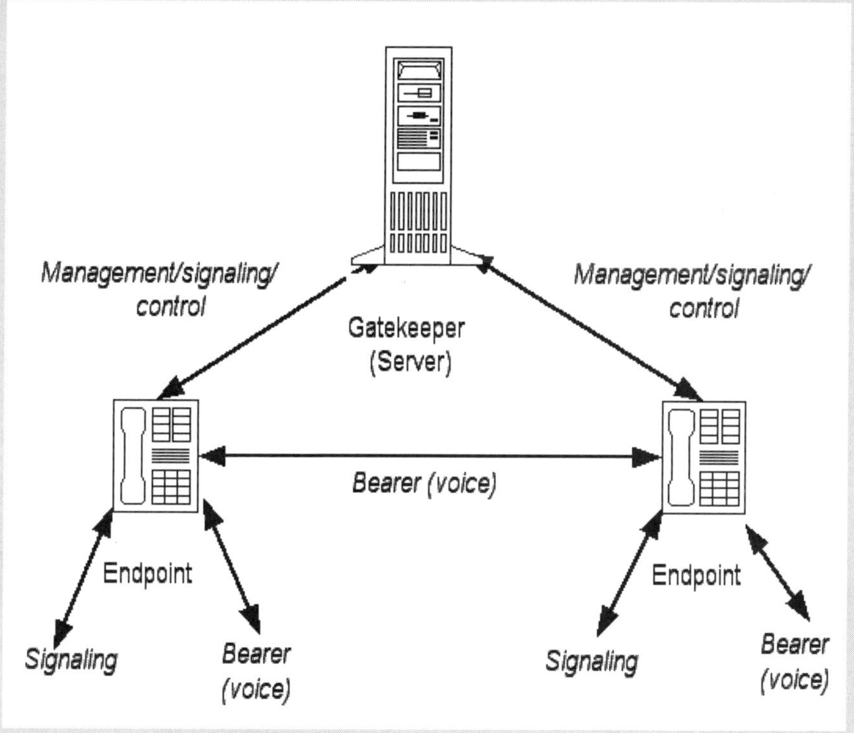

H.323 Endpoint to Endpoint Architecture

Our sincere thanks to Danny Levin, Vice President of Engineering, RADVision, Inc., for this contribution.

Chapter 9

IP-PBX Market Overview

This chapter provides an overview of the leading IP-PBXs on the market today, including the 3Com NBX 100 Communications Server, Cicso Enterprise Telephony System, Ericsson WebSwitch, and Shoreline Communications IP Voice Communications System.

9.1 3Com NBX Communications System

The NBX Communications System first began shipping in March 1998. At that time, the product line was designed and manufactured by privately held NBX Corporation. In March 1999, one year after the introduction of the NBX Communications System, 3Com acquired NBX Corporation. Today, the NBX Communications System is one of the most widely deployed next generation phone systems on the market, with over 4,200 systems and 50,000 phone sets in use.

The NBX Communications System is targeted at small and medium size businesses and branch offices with up to 200 users (3Com has plans to support up to 1000 users in the future). The solution comes in two models:

- NBX 100 Communications System supports up to 200 lines and stations (with a maximum of 100 trunk lines)
- NBX 25 Communications System supports up to 25 lines and stations (with a maximum of 12 trunk lines)

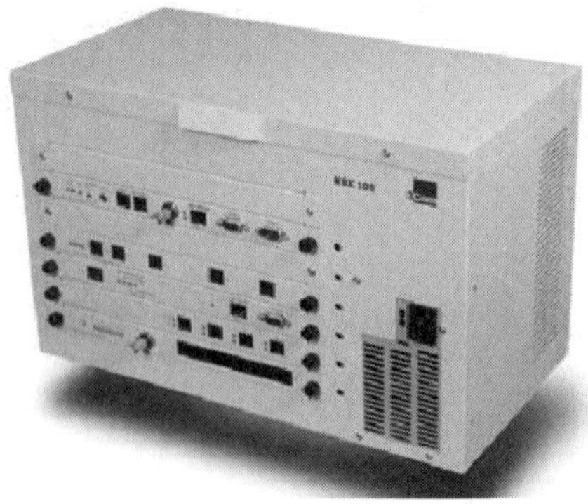

Figure 9-1: NBX 100 Communications System chassis. Copyright (c) 3Com Corporation.

9.1.1 System Components

One of the core components of the NBX Communications System is the system chassis. The chassis houses the network interface cards (voice and data) and the call processing system that manages all incoming and outgoing call traffic. More specifically, the system chassis has 6 slots that can be used for a combination of the following interface cards:

- 4-port analog line card for connecting to analog trunk lines
- 1-port digital line card for connecting a T-1 circuit
- 4-port analog terminal card for connecting analog devices, such as a standard telephone or fax machine
- 8-port 10Base-T hub card for network connectivity and a basic LAN for organizations that do not already have one

As mentioned above, the chassis also contains the call control system — the NBX Call Processor. In addition to managing call traffic, the NBX Call Processor also includes the Advanced PowerMail eXchange (APX) Messaging system. The APX Messaging system provides voice mail services that can be accessed through the conventional telephone interface or from any IMAP4-compatible e-mail client. The NBX Call Processor provides support for up to 12 auto-attendant ports and basic ACD support. An embedded Web server allows the system to be configured from a standard Web browser.

The other integral component of the NBX Communications System is the telephone set. 3Com provides sophisticated, feature-rich desktop phones that include an Ethernet connection for interfacing with the NBX Call Processor over a LAN. These phones offer advanced features, such as, built-in speakerphone, LCD display, Caller ID support, pre-programmed services, 18 programmable keys, volume controls, and more. 3Com also offers an attendant console for managing all calls within an organization from a central location. The attendant console includes Direct Station Select (DSS) and Busy Lamp Field (BLF) status for each extension.

As a pure Ethernet solution, NBX Communications System does not require that the telephones be assigned static IP addresses. In a typical site installation, the phones are considered to be standard Ethernet devices, not nodes on an IP network. 3Com offers an optional add-on product to provide IP addressing for the NBX phones. This add-on is only necessary when using a phone from a remote location. In such a scenario, each remote phone must have an IP address. In addition, internal NBX phones must have IP addresses in order to communicate with the remote users. These IP addresses can be either statically assigned or dynamically assigned from a pool of addresses managed by the NBX system.

Figure 9-2: The NBX 100 Business Telephone. Copyright (c) 3Com Corporation.

9.1.2 The ConneXtions Gateway

3Com offers a software-based H.323 gateway for linking multiple installations together over an IP network. 3Com calls this product the ConneXtions Gateway and it runs on a dedicated Windows NT server.

9.1.3 Summary

The 3Com NBX Communications System is a great solution for small to medium size offices that wish to take advantage of a converged voice and data infrastructure with an integrated, easy-to-use solution.

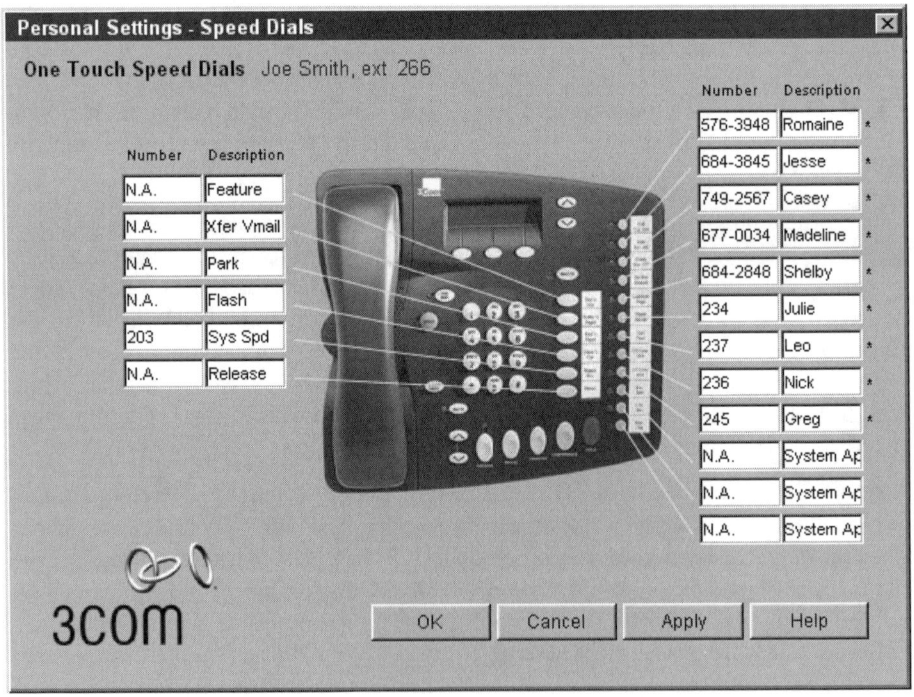

Figure 9-3: 3Com provides PC-based configuration of the NBX phone. Copyright 3Com Corporation.

INSIDER PERSPECTIVE

The 3Com NBX 100 Communications System: Providing High-Performance Telephony to the Small- and Medium-Size Enterprise

By 3Com Corporation

In 3Com's perspective, traditional solutions, like PBXs and key systems, are expensive to install and administer, requiring dedicated wiring infrastructures and either service contracts or an internal staff experienced in telephone technologies. Linking telephones and desktop computers for even basic CTI applications costs hundreds of dollars per phone. Additionally, these traditional telephone solutions are expensive to maintain. Moving, adding, and changing telephones requires skilled service technicians to perform the work. For rapidly growing small- and medium-size firms, which often have continually changing workforces, phone installations are a substantial drain on resources. These systems do not scale well and require firms to install entirely new technologies when growth outpaces their capacity and usefulness.

PC-server telephony solutions, another option, have not gained widespread acceptance because they are cumbersome, difficult, and not sufficiently reliable. They also demand separate wiring for telephones and the phones cease to work should the server fail. This approach also requires that users access their phone services via computer screens, which is for most customers a less convenient and intuitive alternative than simply picking up a handset and dialing a phone.

As a result, although the marketplace offers choices for obtaining voicecommunications, none are tailored to fully meet the budgets of small- andmedium-size organizations. This paper will detail 3Com's answer for this market demand- the NBX 100 Communications System- as well as describe the products architecture, and key features and applications that make this product a viable PBX alternative.

Although the marketplace offers choices for obtaining voice communications, none are tailored to meet the budgets of small- and medium-size organizations. While the reduced cost of computing power has leveled the playing field for data communications—allowing the smallest firms to afford robust desktops—powerful telephony remains limited to larger, resource-rich organizations.

Today's small- and medium-sized organizations demand telephone solutions that are affordable, yet deliver the features of powerful telephone systems, including call handling and even call-center capabilities. Telephony systems for this marketplace must be flexible, easy to scale, allowing for simple administration to manage growth.

3Com's NBX 100 Communications System delivers toll-quality voice communications and the advanced telephony services demanded by businesses today. The NBX system features the sophisticated capabilities of traditional solutions, such as PBXs, but without the costs and complexity. The platform also lowers cost of ownership by permitting users to quickly move, change, and add telephones without any expertise in telephony technologies or reliance on service providers.

The NBX system fully integrates telephony with standards-based Ethernet technology to provide a powerful telephone system that addresses the needs of small- to medium-size businesses and the remote offices of larger enterprises.

Converging Phones with Standards-Based Ethernet Networks

The NBX 100 Communications System is installed as an integral part of a customer's Ethernet-based local area network (LAN). Each NBX telephone is a network device, like any desktop, server, or printer. For organizations without data networks, the NBX system provides a standards-based 10 megabits per second (Mbps) LAN to which additional devices like PCs, servers, and printers can be attached.

The NBX solution unifies voice and data communications, eliminating the need for separate wiring infrastructures for telephones and data LANs. The merging of telephone and data systems permits users to truly integrate voice and data communications for features like unified voice and e-mail messaging and TAPI (Microsoft Telephony Applications Programming Interface)-enabled CTI.

The 3Com voice solution is based on Ethernet, the protocol that has emerged as the worldwide standard for LAN infrastructures due to its simplicity, ease of use, low costs, and scalability. 3Com NBX technology formats audio signals into digital Ethernet packets and forwards them to their destination, where they are formatted back to audio signals. Even over wireless connections, the NBX platform eliminates latency and jitter that compromised the voice quality of early Ethernet telephony. Users can rely on NBX phones for reliable, crystal-clear voice communications.

Each NBX system links to outside phone lines (loop-start analog lines or digital T1), allowing users to call any phone in the world, exactly like traditional telephony. The NBX telephony solution also provides quality phone services over the wide area network (WAN) for toll-bypass communications with other networked sites. Moreover, NBX technology supports H.323 services, enabling economical Voice-over-IP communications. This functionality allows users to make toll-free, long-distance calls over the Internet to branch offices or corporate headquarters.

NBX phone systems have a proven record for reliability. They run independently of both network and PC operating systems and do not require servers or desktop clients. For this reason, even if every PC on the network crashes, the NBX platform will continue to provide fully functional telephony, ensuring the flow of mission-critical voice communications.

Innovative Architecture Delivers Functionality, Simplicity, and Economy

An NBX system uses two prime hardware components—a rack-mounted chassis and multi-line telephone sets.

The chassis connects via standard interfaces to the LAN infrastructure and provides the processing for all call traffic. It supports multiple processing and interface modules, enabling customers to scale phone systems as needed. Each four-port analog line interface card connects up to four public switched telephone network (PSTN) lines with built-in Caller ID support. An NBX Digital Line card permits customers to take advantage of T1 services for high volume, low-cost connectivity to local and long-distance telephone service providers. The system's scalable call processing engine doubles as a voice application server, delivering advanced features like multiple, multi-level auto attendants and CTI.

Distinguishing the NBX system from traditional phone solutions is the simplicity with which the sets connect to the system. Unlike the master/slave paradigm of the PBX, NBX phones, as Ethernet devices, connect directly to any Ethernet port in the LAN. Users simply plug their phones into the network as they would a printer or laptop computer. To relocate phones, they merely unplug them from one location jack and connect them to another. The distributed architecture of the NBX system enables each telephone to retain its identity (i.e., its customized speed dials, usage authorization and extension number). Consequently, NBX solutions enable organizations to arrange and re-arrange their phone system according to prevailing needs with unprecedented ease. This simplicity dramatically reduces the NBX system's cost of ownership over conventional telephone systems, which are difficult and expensive to modify.

NBX 100 architecture is distributed for increased control and flexibility throughout the network. Each NBX telephone contains a digital signal processor (DSP) to handle all incoming and outgoing voice traffic, providing sound quality or QoS at each intelligent end point. The telephone set provides the full range of features, such as speakerphone, conferencing, transfer, redial, caller ID and speed dialing, that users have come to expect from premium phone systems. With one touch, users can instantly access voice messaging services, speed dial, and voice mail. Moreover, achieving truly converged voice and data communications, each phone also features a built-in mini-hub or data jack, allowing users to plug computers into the network via the telephone set.

Management of the entire NBX phone system is simplified by the Web browser-based NBX NetSet(tm) administration utility. This user-friendly GUI (graphical-user interface) program enables employees to customize their phones' feature set, ensuring each is configured for optimum productivity. The intuitive NBX NetSet works with any familiar Web browser, such as Netscape Navigator or Microsoft Internet Explorer. Every NBX system includes the NBX NetSet administration utility for both users and administrators.

Benefits & Features

The NBX system makes advanced telephone solutions affordable and practical. 3Com's LAN-based telephone system delivers superb quality, function-rich business telephony and advanced CTI functionality at significantly competitive costs. It enables organizations to operate more productively, reliably, and cost-effectively. Moreover, the solution delivers exceptional value by providing a wealth of features that otherwise would cost extra and require hardware upgrades and service calls.

Convergence of Voice & Data Communications

The NBX system integrates voice communications with data traffic on a single networking infrastructure, eliminating the separate wiring infrastructure and significantly reducing the expertise and resources needed to maintain these systems.

- Reduced Cost of Deployment—Customers leverage their existing Ethernet network cabling to fulfill all their communication needs. By delivering both voice and data communications over a single wire, the NBX solution greatly reduces deployment costs and resource-draining complexity.
- Lowered Total Cost of Ownership—By allowing for easy moves, adds, and changes of phones, the NBX architecture substantially reduces the cost of ownership. Organizations can deploy full-functioned telephone services without incurring the maintenance and administrative expenses demanded by conventional phone technologies. By reducing the need for costly service contracts and on-site service technicians, even small locations can benefit from the sophisticated telephony features once limited to large corporations.
- Flexibility—The NBX platform operates in shared and switched 10Base-T Ethernet environments commonly used by small and mid-size sites. The system functions in virtually any networking architecture and any WAN technology, including ISDN, ATM, T1, Frame Relay, xDSL, cable modems, and wireless Ethernet links. Optional analog terminal adapter cards support tip/ring or analog devices, permitting devices like fax machines, cordless phones and single line (2500) sets to operate seamlessly on the system. Users that rely heavily on telephones, like call-center agents, can use a headset and the NBX pcXset PC telephone client for hands-free voice communications.
- Reliability—The NBX telephony solution delivers 99.99% reliability, which translates to only one minute of scheduled downtime occurring for every 1,000 operating hours. The NBX system works independently of any operating systems on the network and will continue to perform even if any or all other network devices cease to function. This essential design characteristic means that users and callers will have telephone capability whenever they need it.
- A Free LAN—For sites without a data network, installing an NBX system also provides a built-in LAN, ready to be leveraged for data transmissions. Users can

plug in PCs, printers, servers, and other devices into the network for fully functional connectivity. The NBX architecture delivers the opportunity and simplicity of convergence.

- Standards-based—The NBX platform offers open, standards-based architecture that works with any existing desktop computers, servers and Ethernet switches, routers and hubs. As a result, customers avoid the costs, hassles, and compatibility issues raised by proprietary technologies.

Premium Quality Business Telephone Services

- Call Management Features—To power businesses, every NBX phone system provides a host of productivity features. Each NBX telephone offers speakerphone, conferencing, caller ID, speed dialing, and call pick-up from any other phone. NBX systems deliver call forwarding, built-in system-wide paging, call detail reporting, and user-programmable keys with LCD displays for greater ease-of-use and efficiency.
- Multiple and Multi-Level Auto Attendants—Automated attendant allows users to support or replace receptionists and improve the flow of communications during non-business hours. New enhancements enable NBX users to customize inbound routing by creating up to 99 different automated attendant greetings, each tailored for a division, workgroup, or individual. Using the NBX NetSet administration utility, organizations can provide up to 20 levels of main menus and sub menus that enable callers to route themselves quickly and accurately to desired individuals or workgroups. Every inbound line or DID (Direct Inward Dialing) number can have its own custom greeting (including time-dependent greetings), main menu, and submenu selection choices. As a result, firms can organize call patterns according to exact needs, ensuring customers, clients, and vendors access to their parties around the clock.
- Time-of-Day Service Modes—Firms can further customize NBX systems by programming different automated attendant greetings and call routing paths for various times of day, such as Business hours, Non-business hours, and Lunch. A receptionist can press a button to put the system in Lunch mode at any time, thereby routing callers to a secondary receptionist position.
- Voice Messaging Features—Powerful voice messaging features are standard with every NBX solution. Users can easily access voice mail and deploy personalized greetings, password security, and group messaging. The system even offers phantom mailboxes so workers without phones can receive and manage messages.
- Programmable Call Center Hunt Groups—The NBX platform can route incoming calls to the next available person within a group, such as customer service, technical support, or order entry. For greater flexibility, the technology allows both linear and circular hunt groups — up to 48 groups per system. With linear

hunting, users organized into groups share a common extension. Inbound calls are routed in a sequential, predefined order. In circular hunt groups, calls are routed automatically to the next available member of the group. Calls might go to the next free salesperson within a sales group. If no one is available, the system can stack the callers in a queue or provide alternative coverage. Either way, the hunt group feature ensures that users share the call load and business opportunities are never lost.

- Calling Groups—Calling groups provide coverage for important calls by routing them to a predefined group of extensions until a representative is available. The NBX platform supports up to 48 calling groups to manage call flow, even if an automated attendant answers the phone. If callers, want to speak directly to a person instead of leaving voice mail, they can press "0" on the automated attendant menu, triggering an alert to multiple individuals in a calling group to answer the call promptly.

Powerful communications applications

- Unified Messaging—Every NBX solution includes the APX (Advanced PowerMail eXchange) messaging system, a set of powerful software tools for call answering, routing, and messaging. These tools allow users to seamlessly integrate e-mail and voice mail messaging into a universal in-box for unparalleled ease and efficiency. Employees can view and access voice mail in the same visual way they are used to viewing e-mail, using any IMAP4-compliant contact management application such as Microsoft Outlook Express or Eudora. They can save voice mail as sound files, just as they might save any .wav audio file, and even forward them to anyone in the world with an e-mail address and a multimedia PC. 3Com's Unified Messaging capability allows users to better manage their work, prioritize tasks, and respond rapidly to customer needs. The NBX system ships with 30 minutes of voice mail storage and scales up to 80 hours through software upgrades.
- Local and Off-Site Message Notification and Retrieval—To expedite communications, NBX telephones promptly notify users of new messages via an LED light. Moreover, if users are off-site, the APX messaging system in every NBX platform can notify them of new messages via pagers, telephones, or cellular phones. They then can easily retrieve messages via any touch-tone phone. The NBX solution makes such sophisticated voice communications affordable and practical for even small organizations.
- Call Center Applications—All organizations can benefit from increased contact with customers. The NBX platform delivers advanced call center solutions at no additional cost. Thanks to TAPI-enabled Computer Telephony Integration, the system can recognize each caller and display the appropriate records, such as purchase history and charge card data, on the screen of the next available call center

agent. These time-saving applications, once limited to large, expensive call center operations, are now available to even modest-size firms.

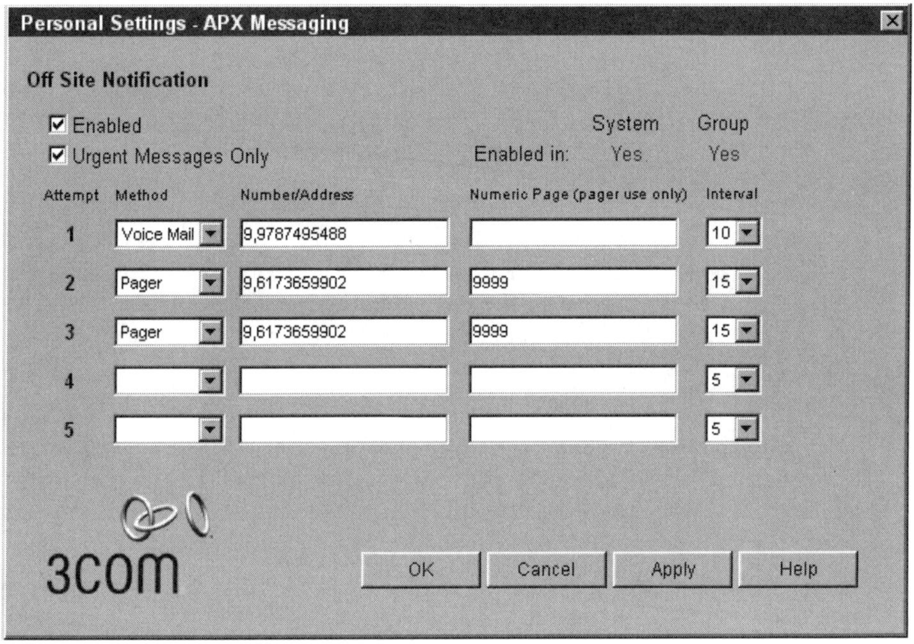

Configuration of the APX messaging system for off-site notification.

Simplified Administration

Despite its robust capabilities and sophisticated features, the NBX system is extraordinarily easy to manage. Organizations no longer have to forego the features and functionality of expensive telephone solutions to control costs and simplify administration.

- Enterprise-wide Management—With the NBX NetSet administration utility, the NBX platform provides administrators and small-business owners with unprecedented control of their phones that is simple and system-wide. From anywhere on the LAN or WAN, they can use the browser-based utility to quickly make changes throughout the enterprise so that NBX phone systems are always up-to-date and ideally configured for maximum productivity.
- Account Codes—For professional organizations that bill clients by the hour, the use of the account code feature of the NBX platform could pay for the entire system. When on the phone with a client, a user enters that client's account code via the telephone set. The system keeps a log of every call by account code, its date and its duration, ensuring accurate accounting of calls and billing. While such a

useful business feature might cost thousands of dollars in other phone systems, it is standard with the NBX solution.

- Dial Plans—The NBX system offers sophisticated dial plans to provide administrators and owners with around-the-clock management of both incoming and outgoing calls. To avoid unauthorized long-distance phone calls that inflate operating costs, dial plans can restrict calls and their routing based on parameters such as time of day, day of week, each user's class of service, and area and country codes.

LAN/WAN/Internet Connectivity]

- Enterprise-wide Communications—In addition to providing users with access to public telephone lines and other phones on the LAN, NBX solutions provide extraordinary connectivity over the WAN and the Internet for greater cost-savings and enterprise-wide communications. Full support for popular connectivity and Quality of Service standards, from PSTN and T1, to 802.1p/Q and IP TOS, provide users with comprehensive deployment options to meet any need.
- Internet Telephony—With its support of the Voice-over-IP H.323 standards, the NBX system offers voice communications over the Internet. Distributed offices can use NBX ConneXtions gateways to place calls to each other over the Internet rather than the public telephone system. As a result, long-distance phone charges can be reduced for substantial cost-savings.
- Supporting Telecommuters—Telecommuters and traveling employees can plug their NBX phones into standard routers at home or elsewhere and link via the WAN or the Internet to the central office LAN and its NBX system. With full access to all of the system's telephony features, workers will seem to be in the office, regardless of their location. Businesses can use this cost-effective feature to attract and retain valued employees, offering location flexibility while maintaining seamless professional communications.

Conclusion

Even as data networks support the worldwide proliferation of e-commerce, the telephone remains critical to organizations, large and small. Capitalizing on this fact, traditional telephony vendors have enhanced telephone capabilities, making phone service crucial to the marketplace. These solutions, however, are proprietary, inflexible, expensive to buy and deploy, and not very expandable. While their functionality provide the large organizations that can afford them with distinct competitive advantages, smaller locations have been left out in the cold.

3Com's NBX 100 Communications System is designed to provide all organizations with reliable, feature-rich telephone systems, but without the costs or complexity associated with traditional solutions. The NBX platform leverages and extends the functionality of scalable Ethernet networks, providing advanced voice communications and the convergence of voice and data for powerful business applications. The NBX system delivers solid benefits because of its many capabilities and the unequalled ease-to-use of these features, thereby ensuring bottom-line cost savings and tangible productivity enhancements.

This Insider Perspective was contributed by 3Com Corporation. Our thanks to Paula Muise of 3Com for providing this submission. Copyright (c) 3Com Corporation.

9.2 Cisco Enterprise Telephony System

The Cisco Enterprise Telephony System is probably the purest IP-PBX solution on the market today. The Cisco solution is an end-to-end IP telephony solution, where all of the phones are assigned static IP addresses and are considered nodes on the IP network. All calls within the system are sent from end-to-end as IP calls. A media/protocol gateway is only used at the interface to the circuit-switched PSTN.

The technology behind the Cisco Enterprise Telephony System was originally developed by Selsius Systems, which was acquired by Cisco in October 1998. Selsius Systems had developed a solution consisting of IP telephones, a Windows NT control system, and a packet-to-circuit-switched gateway for interfacing to the PSTN. In this model, call switching was a service of the network and was separate from the control system. The Cisco solution of today is based on the same basic components and architecture as the original Selsius solution.

Figure 9-4: Cisco IP Phone. Copyright (c) Cisco Systems, Inc.

The heart of the Cisco solution is the Cisco CallManager. The CallManager is an integrated suite of software applications for system setup, configuration, and enhanced features. One of the most significant enhancements in the latest version of the CallManager (version 3) is the ability to cluster multiple CallManager servers and manage them as a single entity. By clustering the servers, the configuration from the master CallManager database is automatically replicated to all of the CallManager servers in the cluster. A single CallManager cluster can support up to 10,000 users and clusters can be linked together for even greater scalability.

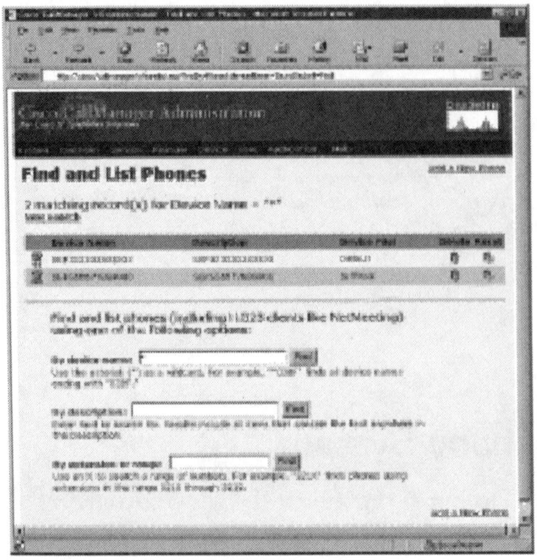

Figure 9-5: The browser-based interface to the Cisco CallManager. Copyright (c) Cisco Systems, Inc.

The Cisco Enterprise Telephony System is an excellent voice communications infrastructure solution for organizations interested in a robust, scalable IP-PBX.

INSIDER PERSPECTIVE

The IP-PBX

By Dave Corley—Product Line Manager, Cisco Systems

Available technology, cost and public policy have heavily influenced the history of enterprise telephony. The early concept of universal service provision promoted by public telephone service providers was a boon to the creation of standards and ubiquitous deployment of service. On the other hand, they fostered monopolistic policies that prevented an enterprise from taking competitive advantage of differentiated services in their business. A more laissez faire public telephony service policy and the advent of electronic computers and digital technologies acted together to provide a foundation for PBX systems that offered enterprise users and administrators the ability to deploy call centers, supplementary user services and management efficiencies. The historical trend continues with the introduction of the IP-PBX.

The Move Toward IP

PC-based and monolithic TDM (time division multiplex) PBXs are built upon a network that is separate from the corporate data network. Until the late 1980s, there was no commercially available means by which voice from TDM-based systems could be converted to packets and transported over a packet or cell-based network. VocalTec and other vendors introduced the VoIP (Voice over IP) gateway to provide such a capability. The early focus of enterprise and service provider IP telephony was founded upon using these gateways to provide an alternate path to the PSTN for long-haul telephone traffic. Enterprises could benefit from the ability to bypass long-haul toll charges imposed by their telephony service providers. The service providers, initially averse to the idea of toll bypass, have since embraced the idea of IP telephony as a means to offer differentiated services to residences and businesses. The utility of toll bypass applications is still apparent, even with the lowering cost of long-haul telephony service. But its usefulness to the enterprise user at the desktop is limited at best. The IP-PBX extends the domain of IP telephony from toll bypass to end user services in the enterprise. Next generation service extension to residential users will follow as price-performance justifies the application.

Under the Hood

The IP-PBX architecture differs substantially from that of traditional TDM designs such as monolithic PBXs and PC-based PBXs. Where the *VoIP-enabled* PBX is a composite of a monolithic architecture, with add-on VoIP gateways, the *IP-PBX* goes the final step—all system devices are fully IP-enabled. The only TDM component in this network architecture is that required for interconnection with legacy TDM networks and devices. The premise for IP-PBX design is convergence of services and infrastructure onto a single, ubiquitous network. As data devices, IP-PBX components are more likely to follow the time-cost models of data networking devices where per-port costs drop at rates of as much as twenty percent per year. Also, the logical proximity of the Web and other IP services to IP-enabled phones, devices and applications allows direct extension of those services to those components. In contrast, TDM-based phones do not have direct access to IP services, relying upon a combination of telephony APIs and a specialized application to extend a service to an associated personal computer. This level of integration is difficult, costly and a pain to administer. The bottom line to financial planners for IP-PBXs is lower total cost of ownership. The bottom line to users is advanced telephony services delivery to what, in the past, has been a one dimensional, voice-only phone system.

IP-PBX architecture de-couples the link between the intelligence of the phone system and network transport. Telephony therefore becomes another of many services offered on the data network. An advantage of such de-coupling and network abstraction is that system components may be distributed anywhere on that network. Also, other media types such as video and collaborative data may be carried over the same network.

Although TDM networks may be configured to carry video and collaborative data, the exercise is time-consuming, expensive and difficult to install and maintain.

PC-PBXs and their larger siblings—monolithic TDM PBXs—are similar in their architectures. The network over which media (voice) is passed on those networks is one-dimensional in that only voice traffic is carried. Further, the nature of the hardware dependency of the architecture limits the ability to scale and distribute to larger networks without considerable cost and effort. Finally, direct access to the World Wide Web, the coming flood of XML- based applications and IP-based video is offered to user terminals such as IP phones and software phones. The result is that IP allows the phone to evolve to a device that offers multiple tailored user services.

A simplified description of a distributed IP-based PBX is presented in the figure below. Voice and signaling traffic is carried on the underlying transport network This network is overlaid with Internet Protocols (IP). The physical and link-layer is abstracted from the system components by IP. On top of this abstracted transport are end-user devices (wired and wireless IP phones, software-based telephones), trunk access devices (VoIP gateways), voice applications (voice messaging, interactive voice response, automated attendant, call centers) and call control/call processing servers.

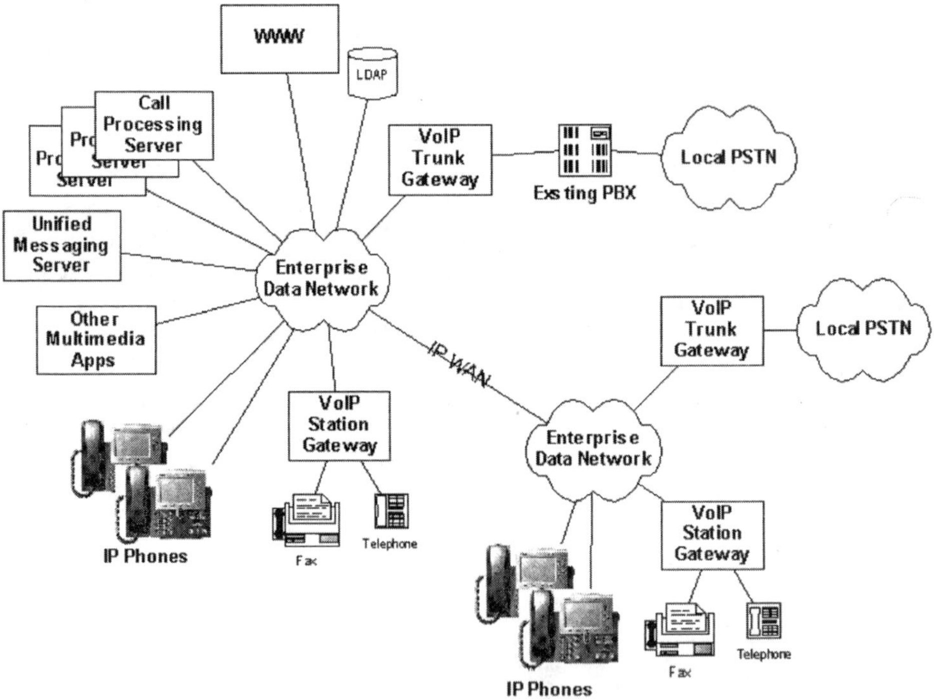

Generic IP-PBX Architecture

Application servers are differentiated from their TDM-equivalents by the method by which base telephony operations are performed. CPU performance on typical server-class PCs has risen dramatically over the course of the past fifteen years. As a result, operations such as DTMF detection and voice streaming, which required dedicated hardware processing in the 1990's may be accomplished in software on off-the-shelf server-class PCs. As an example, hundreds of G.711 PCM streams may be conferenced together on a commercial off-the-shelf server. The PCM-to-linear decode, linear summing, and linear to PCM encode and streaming operation requires only a few MIPs of processing power from the host CPU. Today's server-class PCs are becoming more-and-more capable of performing tasks that have traditionally been performed by dedicated hardware-based signal processors. The removal of another dependency on network hardware promotes distributability, scalability and affordability. The need to perform high-density, complex signal processing operations such as phonetic speech recognition and low-bandwidth encode/decode operations still dictate the need for affordable hardware-based digital signal processor (DSP) farms. But as server prices continue to drop, even these operations are being performed in relatively high density on server-class PCs.

The position of call control and call processing in the IP-PBX depends on the architectural design of specific products. However, three general classes of call control architecture can be assigned:

- Monolithic. All call control and call-processing functions are performed by a single processor. The processor is the master to all slave devices. These slave devices are "stupid" in that they maintain little or no call or configuration state. Slaves include IP phones, trunk gateways, applications and media processing devices. In a purely monolithic design there is no provision for redundancy at the call processing/call control server.
- Device-based. Call control functions are located at intelligent endpoints such as phones, gateways and applications. Multimedia and other services are delivered from intelligent servers located on the network.
- Hybrid. Call control is performed by clusters of intelligent servers. Endpoint devices are largely stupid with respect to call control.

Distributed call control has several advantages including redundancy and load sharing. As an example, consider the case of a hybrid system implementation composed of IP telephones and a network of five call processing/call control servers. The described hybrid call-processing network is the Cisco Enterprise Telephony System—a commercially available IP-PBX developed by Cisco Systems. All IP phones are configured such that, at start-up, the phone downloads a configuration file that lists the order in which the phone would attempt registration to the call-processing server. Once the phone is registered, that server becomes the "active" telephony server for that phone. Because each phone's configuration file may have its own ordered list of call control servers, configuration files may be configured to provide load sharing among a cluster of call processing servers. Intra-cluster call control signaling provides feature transparency among all phones registered in the server cluster. Abstraction of the configuration database allows single point administration for device moves, adds and changes.

Multiple ordered entries in the phones configuration file allows call processing redundancy in the event of failure of a phone's active call processing server. When the connection-oriented signaling link from the active server to the phone is disrupted, the phone detects the disruption and automatically registers to the second call-processing server in its configuration file.

Another IP-based architecture gaining favor is based upon a lightweight peer-to-peer protocol—the Session Interface Protocol (SIP). In this model, call control intelligence is provided in the terminal (gateway, phone, and application). Network servers provide services for directory, forwarding etc. Because the network devices are fully distributed, this type of network is highly redundant.

How IP-PBXs are Deployed Today

As of 2000, the IP-PBX is moving from a collection of disparate technologies deployed by early adopters, to a composite solution of phones, telephony servers, applications and network services in numerous large-scale enterprises. The inherent scalability and interoperability of the solution allows customers to attach an IP-PBX solution with as few as a single phone and a trunk gateway to an existing TDM PBX thereby extending the TDM PBX. By far, this is the model being deployed as late as mid-1999. A significant number of users are deploying IP-PBXs as large adjuncts to their existing PBX, awaiting TDM lease expiration or deploying to take advantage of distributability over existing QoS-enabled IP networks. A smaller number of users are deploying in green-field environments where new buildings are being populated or complete telephony system renovation is in order. In this model, the entire voice, video and integrated data infrastructure is converged.

Challenges

As IP-PBX solutions mature beyond the early adopter phase, many challenges face large-scale deployments. These challenges include:

- Integration with legacy TDM applications such as voice mail and call centers represents a substantial challenge.
- Voice mail inter-networking. While standards for inter-networking voice mail systems such as AMIS and VPIM exist today these standards are not widely adopted by legacy market leaders.

- Enterprise IP networks may be widely distributed among remote sites. This represents a unique conceptual advantage of IP-PBXs in that they can be conceptually deployed as a single, manageable PBX among multiple sites. The challenge is provision of adequate quality of service measures in network products and configuration. QoS protocols such as DiffServe, VLANs and RSVP (Resource reserVation Protocol) show promise of delivering QoS and allowing realization of the promise of a single IP-PBX network.
- Interoperability. The promise of interoperability is the replacement of one vendor's products for another vendor's best of breed products. The challenge this represents is implementation of the standard by vendors concerned as well as sufficient interoperability testing.
- Extensibility. Vendors of telephony adjunct products such as voice/unified messaging systems, multimedia call centers have traditionally developed their products to interoperate with legacy TDM PBXs. Integration with PBX call processing is through a variety of APIs, including TAPI, TSAPI, CSTA and JTAPI. Transitioning these products to the IP world represents a challenge, but one that leading edge vendors are moving towards. These vendors recognize the promise of integrating IP-based services with their existing products in a more efficient manner than TDM-based solutions afford.

Summary

The IP-PBX represents the second phase in the evolution of Voice over IP products. Where the first phase products focused on the ability to save money through bypassing long-haul toll charges, the second phase—manifested by the IP-PBX—offers convergence of a rich set of existing IP services with those of the traditional telephone systems. The synergy of this approach offers business differentiation and cost efficiencies to managers of corporate telecommunications systems. The IP-PBX is here to stay. Its application in corporate and other environments will revolutionize the way we do business.

This Insider Perspective was contributed by Dave Corley—Product Line Manager for Cisco Systems' Enterprise Voice Unit. Dave previously served in technical and marketing positions with Selsius Systems, Incite and Voice Control Systems. He is a former nuclear-trained submariner serving on three nuclear submarines. He received his Bachelor's degree in Aerospace Engineering from the U. S. Naval Academy and a Master's degree in Mechanical Engineering from the U. S. Naval Postgraduate School.

9.3 Shoreline Communications IP Voice Communications System

The Shoreline Communications (formerly Shoreline Teleworks) IP Voice Communications Systems is a unique, distributed IP-PBX designed to provide seamless scalability across one or more sites connected by an IP-based LAN or WAN. Shoreline's solution is based on an architecture referred to as DIVA (Distributed Internet Voice Architecture), which is designed to distribute the intelligence and call control capabilities throughout the network. At present, a solution based on the Shoreline IP Voice Communications System can connect up to 1,200 local and remote users in a single integrated solution.

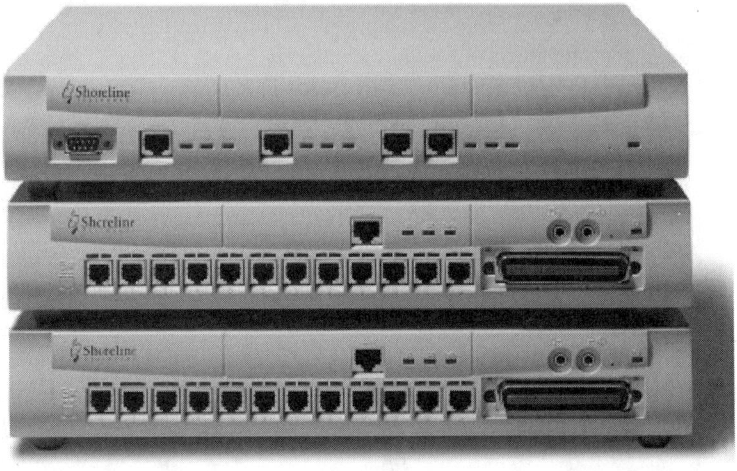

Figure 9-6: Shoreline Communications' stackable voice switches. Copyright (c) Shoreline Communications, Inc.

In contrast to some of the other solutions discussed in this section, Shoreline does not currently provide packet-switched voice directly to the desktop. Shoreline uses the IP data network as an expansion bus interface between multiple voice switches. The data network is also used to distribute call management and routing intelligence to all of the voice switches on the network to create a single system image across all locations.

9.3.1 System Components

The core components of Shoreline's distributed solution are:

- ShoreGear Voice Switches—The voice switches provide connectivity to the PSTN, desktop phones, and the data network. The PSTN interface can be either analog (12 or 24 ports) or digital T-1. The interface to desktop telephones is a standard analog interface. The LAN connection is an auto-sensing 10/100 Ethernet port, which is used to link multiple voice switches over a corporate network. Shoreline also offers a 4-port switch designed specifically for remote workers with DSL service. The beauty of this switch is that it is fully integrated with the other switches in the enterprise, giving a remote employee the same feature set as an on-site employee.

- ShoreWare Director Software—The ShoreWare Director software is a Web browser-based application that is used to set up users, trunks, and switches. As a browser-based application, the ShoreWare Director can be run from any PC on the network, allowing system administrators to configure a distributed Shoreline solution from virtually anywhere. When a new voice switch is added to the network, the ShoreWare Director automatically recognizes the switch and adds it to the existing system, making scalability seamless and easy.
- ShoreWare Voice Services Software—The ShoreWare Voice Services software is a Windows NT application suite that provides advanced phone system features, such as voice mail, auto-attendant, ACD, and call routing across the network. The software also provides complete Call Detail Recording, which is stored in a standard SQL database.

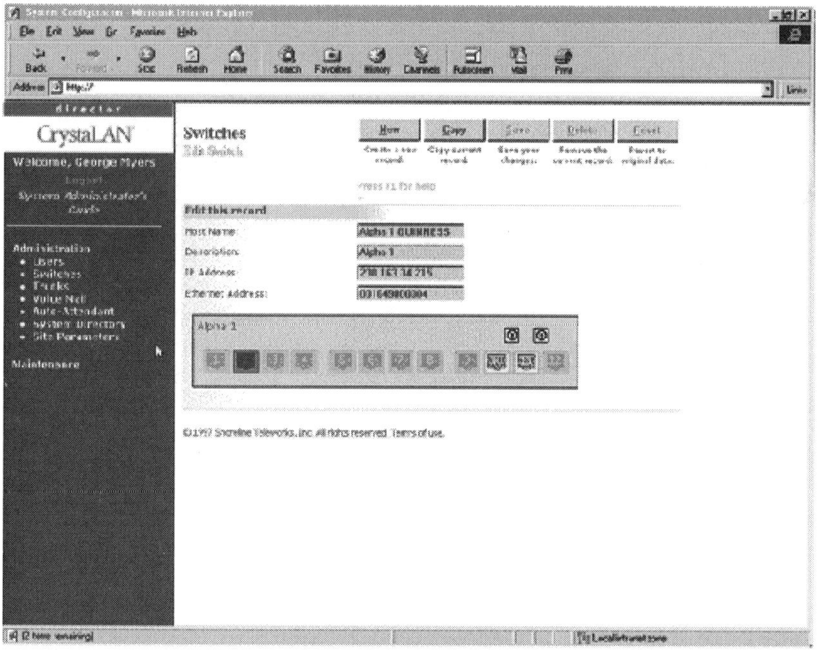

Figure 9-7: The ShoreWare Director is a browser-based application for configuring users, trunks and switches. Copyright (c) Shoreline Communications.

- ShoreWare Call Manager Software—There are three different versions of the ShoreWare Call Manager software: a personal version, a workgroup version, and an operator version. Each is designed to provide a different level of PC-based call control to the user. These software packages give the user point-and-click access to all of the phone system features. In addition, the personal version provides a visual interface to voice mail and the ability to set up personalized call-handling rules. The operator version is specifically designed for a receptionist or anyone who professionally manages calls for others.

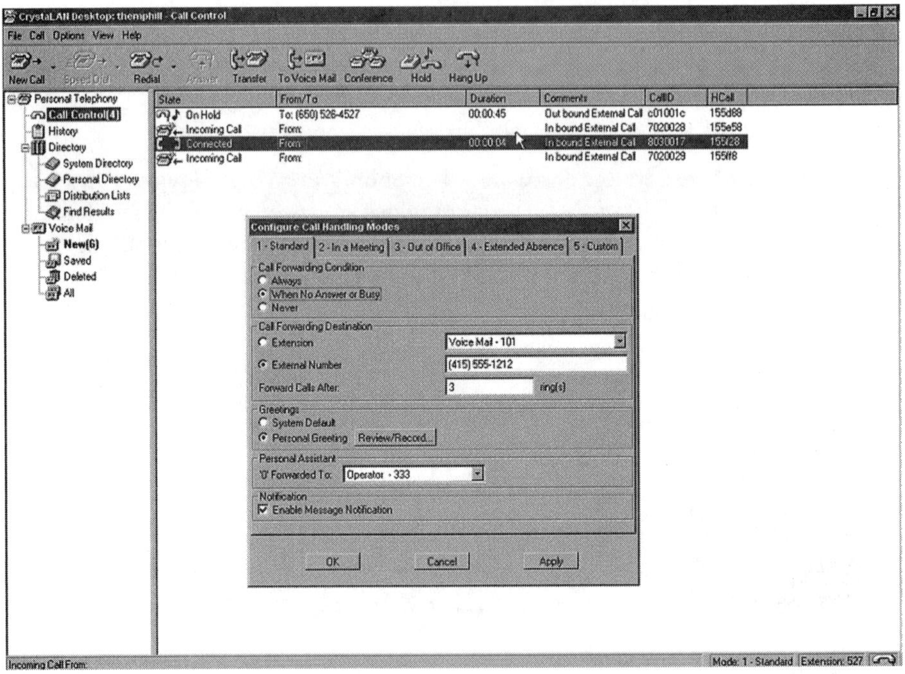

Figure 9-8: The ShoreWare Personal Call Manager provides point-and-click access to all phone system features. Copyright (c) Shoreline Communications.

9.3.2 Summary

Shoreline Communications offers an exciting solution for organizations with multiple networked locations that wish to implement a single, integrated voice communication solution. By using packet-switched voice as the expansion bus interface between voice switches, Shoreline optimizes the value of the packet-switched network without the need to deploy more costly IP-based desktop phones. In addition, the Shoreline Teleworker Voice Switch is one of the most effective solutions for telecommuters and remote workers. With scalability up to 1,200 ports, many organizations will find the Shoreline IP Voice Communications System to be a great fit.

9.4 Ericsson WebSwitch

Do you need more evidence that computers, networking, and telephony are merging? If so, take a good look at the Ericsson WebSwitch. Just the appearance of the WebSwitch is enough to make you realize that it's more than a traditional phone system. At first glance, you might even mistake the WebSwitch for a network appliance, such as a router or hub. In fact, the Ericsson WebSwitch was one of the first *hybrid* phone systems on the market- offering both circuit-switched and packet-switched telephony to the desktop, while supporting traditional circuit-switched trunks and using IP telephony to provide a scalable, distributed platform for enterprise-wide voice communication.

Figure 9-9: The Ericsson WebSwitch 2000. Copyright (c) Ericsson

The WebSwitch was introduced to the market in 1998 by TouchWave, Inc. In 1999, Ericsson acquired TouchWave and the WebSwitch product line. The WebSwitch is available in configurations that support up to 32 or 64 analog extensions and analog or digital (T-1 or E-1) trunk lines. With built-in voice mail, auto attendant, music-on-hold, and graphical administration / desktop CTI applications, a standalone WebSwitch unit is a feature-rich phone system for any small office. What makes the WebSwitch unique is the existence of an Ethernet port on the front of the unit and an integrated H.323 (v2) compliant IP telephony gateway. When you connect the WebSwitch to your IP-based LAN or WAN, you establish the infrastructure that allows multiple WebSwitch units to acts as one large super system, using the network as a virtual expansion bus. The built-in gateway also provides support for H.323 PC applications, such as Microsoft Windows NetMeeting and IP telephones, such as Symbol Technologies' NetVision wireless LAN phones.

Multiple WebSwitch units connected to a LAN or WAN will route calls between the units using IP telephony. Inside each WebSwitch unit is an IP telephony gateway that will convert a circuit-switched call from an outside trunk line or internal extension to a packet-switched call that can be transmitted over the IP network between the units. Centralized administrative and directory services allow a multi-unit system to be managed as one large system.

To put this in more concrete terms, if a user were to call another user connected to the same WebSwitch, the call would be handled as a regular circuit-switched call. However, if the user were calling another user connected to another WebSwitch, the call would be transported over the LAN or WAN between the WebSwitch units. All of this happens in a way that is completely transparent to the user. The WebSwitch units communicate with one another to provide users with a global, enterprise-wide dialing plan. Although each WebSwitch is an independent device, they work together to function like one large system.

9.4.1 Distributed Telephony

The idea of using a LAN or WAN as the backbone for a distributed phone system has a number of advantages over the traditional centralized phone system architecture, including:

- No central point of failure—If a unit fails only part of the overall system will lose service. Additionally, since each unit is relatively inexpensive and easy to replace, therefore keeping a spare unit on-hand is practical for emergency situations.
- Reduced long distance tolls—Geographically distributed offices can reduce long distance tolls by using their existing data communication infrastructure for voice communication.
- Scalability—Increasing the capacity of the phone system can be accomplished by adding another switch to the network. With the right data network infrastructure in place, in theory, the capacity of such a system is unlimited.
- Enterprise Telephony—Organizations will be able to easily create a single large system across multiple locations. There will be one enterprise-wide directory that makes it as easy to dial a remote co-worker as it is to dial a colleague in the next office.
- Simplified Installation and Wiring—Each switch can be located near the users that it serves to simplify installation. By plugging a switch into the data network it is integrated with the rest of the system with no additional wiring—this is especially useful for multi-building sites or "campus" settings.
- Graphical Administration—By using a PC connected to the LAN, you can setup and configure the system with an easy-to-use graphical user interface.

The Ericsson WebSwitch, with a built-in IP telephony gateway, was one of the first distributed telephony solutions on the market when it was launched in 1998. Although there are still quality of service (QoS) issues with IP telephony over the unmanaged Internet, IP telephony over a managed corporate network is practical.

9.4.2 Test Drive

Unlike many of the PC-based systems that we've tested recently, the Ericsson WebSwitch is a turnkey appliance that is almost ready-to-run right from the box. There are no expansion cards to install or drivers to configure, making the WebSwitch one of the easiest phone systems to set up.

The first step is to simply plug in the trunk lines, the extensions, and the power supply. The installation guide then describes how to configure an RS-232 connection to the WebSwitch using Windows HyperTerminal and an included RS-232 cable. With this connection you can access the command line of the embedded operating system in the WebSwitch to set basic system parameters such as the date/time and the IP address of the unit. The documentation describes how to change the IP address to one that is suitable for your network.

After selecting an IP address, plugging an Ethernet cable into the unit, and rebooting it (by disconnecting and reconnecting the power supply), you are ready to configure the WebSwitch with SwitchLink, a Windows-based software program. The software is shipped on a CD-ROM and can be installed on any PC connected to your LAN. SwitchLink reads configuration information from the WebSwitch and allows you to view and edit the configuration.

The SwitchLink's graphical user interface of is extremely intuitive. SwitchLink presents a tree-view list of the configuration settings of the WebSwitch. This list appears on the left side of the screen and the options for each item appear on the right side of the screen. Simply click on a section of the configuration tree to view or edit the options. For example, trunk lines can be configured to route all calls to the auto-attendant greeting or to a specific extension. Setting up extensions requires assigning names and passwords. The WebSwitch also supports a basic hunt group or call distribution feature, which can be configured with SwitchLink.

After making configuration changes with SwitchLink and recording an auto-attendant greeting from an extension connected to the WebSwitch, the changes can be downloaded and when the WebSwitch is rebooted, the changes become active.

Ericsson also includes a handsome graphical user interface for performing desktop call control. This software, called PhoneLink, provides standard call control features (dialing, transferring, placing a call on hold, etc.) as well as a built-in address book for quick access to frequently used phone numbers. PhoneLink can also be used to configure call-forwarding preferences so that callers can locate you if you are away from your desk or out of the office.

The basic set up of the WebSwitch is extremely straightforward and does not require any previous telephony experience.

9.4.3 Summary

The WebSwitch provides an excellent compliment between traditional circuit-switch telephony and IP telephony. By supporting analog extensions and trunks lines, the WebSwitch can easily integrate into any office using existing telephones and phone lines. With a built-in IP telephony gateway and software to manage a distributed network of units, Ericsson has made it easy to build a fault-tolerant, enterprise system that truly appears as a single unified system.

Chapter 10

The Virtual Phone System

10.1 Introduction

So far, we've explored a range of different types of phone systems - from traditional PBXs to IP-PBXs. All of these solutions have one thing in common: they require an on-premise installation. Yet, there are circumstances where an on-premise installation is either impractical or overkill. Fortunately, telecom de-regulation, advanced computer telephony technology, and a dynamic business environment have created a market for virtual phone systems.

A virtual phone system provides advanced PBX-like features without the need for on-premise equipment. Advanced computer telephony technology has allowed vendors to develop robust and full-function telecom applications. In turn, telecom deregulation has allowed telecom service providers to offer these applications to the consumer. Virtual phone systems not only provide PBX features to businesses, but also offer other sophisticated call completion services for individuals.

In this chapter we will cover some systems currently in use, ranging from full virtual PBXs (VirtualPBX and Lynx Office) to universal number and telecom assistants (Lynx Connect, Webley, Wildfire) to unified messaging suites (ThinkLink). But first, let's look at an overview of the market.

10.2 Market Trends

Phone systems allow businesses to efficiently handle and route calls. Yet, the expense of installing and maintaining any type of PBX becomes higher (on a per user basis) as the business gets smaller. Virtual phone systems can provide inexpensive call management functions at per user cost and scale from one to 100 users with little trouble.

These services are ideal for small organizations, geographically distributed companies with multiple sites, mobile workers, and telecommuters.

Some of the most significant trends in the virtual phone system market include:

- Single-User vs. Multi-User Systems—Most vendors of virtual phone systems have designed their solution for individuals or businesses, but not both.
  - Single Person Virtual Phone Systems—A few telecom service providers cater to the needs of the single professional. The services provide a single number (universal number) from which fax and voice calls are properly routed. The more upscale of the services provide virtual assistants to notify users of caller information and provide speech recognition to interface with the phone system. Some universal number services also provide a Web-based interface to manage system features and e-mail. All of these services seek to simplify calling the user (one number) and aid the user in maintaining sophisticated call features (such as conference calling) with any type of available phone.
  - Multi-Person Virtual Phone Systems—These services provide single number access to an organization and its employees. Calls are answered by an auto-attendant and the caller can select a specific department or extension. To the caller, these virtual office phone systems are indistinguishable from on-premise PBX solutions. From a user's perspective these systems typically offer a universal number to access the system for retrieving voice mail, setting up preferences, or placing calls. For example, calling a co-worker is as easy as dialing into the system and entering the co-workers extension. The more sophisticated solutions also offer support for virtual ACD workgroups. No matter where employees are located, they can participate in an ACD workgroup—a great solution for distributed organizations! Again, from the callers perspective the fact that these employees may be geographically distributed is not apparent. In addition, these systems can grow with the business and usually the costs scale.
- Web-based Control—Although the feature set of most virtual phone systems probably won't change much in the next few years, we expect all of these systems to offer Web-based interfaces for administration and configuration. Web-based interfaces make virtual phone system significantly easier to manage—especially when it comes to setting up complex features such as an ACD workgroup.

  Also, we expect to see better integration of current call management options with Web-based applications. For example, vendors are beginning to offer visual Web-based interfaces for the management of conference calls—a great improvement over touch-tone control.

- The Application Service Provider (ASP) Phenomenon—The concept of outsourcing computer and networking services is becoming increasingly popular. ASPs are now offering a wide range of business applications—from office productivity suites to databases and accounting systems. As this paradigm gains further acceptance, we expect businesses to become more comfortable with the idea of outsourcing their voice communications solutions as well.

iPBX Hosting: The Virtual Phone System Meets the IP-PBX

What happens when you take the concept of the virtual phone system and cross it with the IP-PBX? The result, according to Netergy Networks, is the iPBX—a unique solution that provides small and growing businesses the flexibility of Centrex, the benefits of convergence, and the advanced functionality of an IP-PBX.

The concept is simple—and, again, it's all about convergence. Today, most companies have a provider of data services (an ISP) and a provider of voice services (a LEC). However, with the convergence of voice and data, having two separate services providers will soon be considered redundant (of course, there is some value in having a redundant configuration from a fault-tolerance perspective, but that's a different topic).

Since voice can be packetized and treated like data, why not consolidate all voice and data communication into a single broadband network connection, such as T-1, DSL, or cable? Furthermore, since IP-PBXs allow the call control component of a phone system to be located anywhere on the network, why not host it at a service provider's facility and take advantage of 24x7 administration, backup, etc. In such a configuration, the only on-premise components are the telephones—either IP phones or standard analog phones connected to a media gateway. And, of course, all system management can be done remotely with Web browser-based applications.

At least one company, Netergy Networks, is addressing the market for hosted, IP-PBX solutions with its Netergy Advanced Telephony System (ATS). The Netergy ATS includes an iPBX and the software to allow service providers to host multiple iPBXs on one or more servers.

As mentioned throughout this book, what makes such a configuration possible is the inherent modularity of an IP-PBX. With an IP-PBX, the various components of a phone system are distributed throughout the network. The IP network provides both a signaling (i.e. call control) and media path between the system components. With such an architecture the various components of a solution can be located where it is most appropriate—on-premise or at a service provider.

10.3 Advantages and Disadvantages

The main advantage of a virtual phone system is that a user saves the expense and time of installing and maintaining equipment in-house. Small or mobile offices may not be able to afford an in-house system, and in many cases they may not have the technical expertise to maintain the system. For a monthly fee and a usage fee (usually per minute), a company can have a sophisticated phone system without the hassle of maintaining or finding space for it.

Another advantage of a virtual phone system is that they are well suited to mobile users. There is no need to set up a device or leave the PC turned on at the home office (if there is one). With most virtual phone systems, the features are controlled over the phone, which means the system can be controlled from anywhere.

These two advantages also turn out to be a disadvantage. For example, controlling the complicated phone system over a touch-tone phone can be frustrating. Fortunately, some vendors have provided Web-based interfaces to these phone systems.

Finally, the cost of using the services can be a major disadvantage. It seems that the rate of adoption of these different services is related to the per-minute charges. The cost structure is out of sync with the general trend towards a "bucket of minutes" associated with highly competitive wireless services or local services, or very low per minute rates of local and long-distance services. Many of the services started before or soon after deregulation, and still maintain the high usage rates of that time.

The benefits of virtual phone systems must be closely analyzed relative to the cost of using such a system. Different from the large onetime charge with occasional upgrade fees typical of a traditional PBX, virtual phone systems carry variable usage fees and recurring monthly fees. These variable usage fees can often be difficult to control and manage.

Finally, a major consideration in selecting a virtual phone system is the feature set and ease with which the feature set can be accessed and configured. Are features easy to manage? Is the programming interface (Web or phone) cumbersome? And, even though the equipment is not located on the user's premise, knowing where it is located is important. Location in this market can have a significant effect on the cost of use. If the virtual phone system is not located within your local calling area, you may be paying long-distance charges whenever you access voice mail, dial into the system to place calls, etc.

Chapter 11

Virtual Phone System Market Overview

The virtual phone systems selected for this overview are among the most prominent in the market. We also chose vendors that showed promising or well-designed services that included some sort of virtual phone system feature, such as call management. The vendors discussed in this overview include: Linx Communications, Netergy Networks, ThinkLink, VirtualPBX.com, Webley Systems, and Wildfire Communications. This overview covers the history, market, and features of the different services.

11.1 Linx Communications

Linx Communications, Inc. was founded in 1996 as a communications service provider. It arose from the observation that call routing and call completion could be simplified and outsourced for businesses and mobile workers. The intention was to design a system where users have access to all calls, messages, and faxes via a local call. Linx is a leading service provider of one-number call management and routing. Linx has recently added a Web-based interface as an additional method for users to manage their accounts. It is currently working on adding e-mail capabilities.

Linx Communications is focused on providing unified communications services to the mobile professional, telecommuter and small office/home office markets. They sell direct and through resellers, such as CLECs, ISPs, value-added resellers, other service providers.

11.1.1 Features—LinxConnect

This discussion of LinxConnect features is based on actual use. When you sign up for a LinxConnect number you are given a single personal number. You then call the number and set up the account for the first time; setting passcodes, and recording various greetings. Currently, you can get a local number in Boston, Washington, D.C.,

San Francisco, Los Angeles, and New York City. More cities are being added all the time. Alternately, you can get an 800 number, but unlike the local number, the users must pay for incoming calls.

From this personal number (local or 800) one can manage calls, receive voice and fax messages, manage account details, record greetings, and reroute faxes. LinxConnect is not just for the consumer; however, larger corporations may not want to change the corporate voice mail system in order to user Linx. In these cases, Linx Communications can integrate the LinxConnect voice mail system with the corporate voice mail system. For example, 3Com Corporation has integrated their corporate messaging systems with Linx services. And the LinxConnect system can also be used to call other subscribers, making the system similar to an internal corporate directory system. Linx Communications also has a virtual PBX product called LinxOffice (more below).

The system is easy to use and comes with many options. For example, this is what happens when a caller dials your LinxConnect number:

1. They hear your personalized greeting. You can have different greetings stored in the system for when you are unavailable or available. You can also store a third temporary greeting that is special. This keeps you from having to overwrite your greetings when your availability status changes.
2. The caller is given the option of contacting the user or leaving a message.
3. If the caller attempts to contact the user, the caller is prompted for their name.
4. Linx attempts to complete the call by simultaneously trying up to three numbers the user has designated as "follow-me" numbers. This is faster than attempting three numbers in succession.
5. If someone at any one of the three numbers picks up the phone, the Linx system provides an introduction: " ...is holding for..."
6. The user has the option of hanging up, accepting the call, sending the call to voice mail or another number, or letting the caller join a conference call.
7. If the call is not completed, the caller will be asked to leave a message or use other options, such as a voice-annotated fax. The user can actually listen to the message as it is being recorded, much like a standard answering machine.

If you want, you can have the Linx system page you. The page message will have important information regarding the nature of new calls, faxes, and voice mail. The Linx system can now notify wireless phones via e-mail.

When the user is connected to a caller or connected to the Linx system, there are a bevy of call management features the user can take advantage of. For example, the user can listen to the message as it is being recorded, or put a caller on hold and make another call. Remember, it is actually the Linx system that is managing the conversation. If, when you are in a call received through the Linx system, and someone calls you, you will hear a tone, similar to call waiting. You can then terminate the current call, or place it on hold, and then manage the new call as usual.

The LinxConnect account phone menu system seems daunting, but the menu system is not too difficult to navigate, and there is a user manual to help you get an overview of all the features of the account. Over the phone you can:

- Listen and manage messages, such as deletion and forwarding, or return calls (phone numbers are associated with messages via caller ID or caller input)
- Manage, retrieve, or reroute faxes to other numbers
- Set personal options, such as passcode, availability, greetings, pager settings
- Set follow-me numbers
- Contact other subscribers directly or through a directory
- Manage calls
- Set up conference calls
- Set up speed dial numbers (speed dial numbers are only available when logged into the system)
- Set up a personal directory, including distribution lists (also, only available when logged into the system)

Indeed, the LinxConnect user account has a huge variety of settings. Simple management of faxes and voice mail is easy, although there can be some confusion when changing settings and accessing advanced features. Fortunately, Linx Communications is rolling out a Web-based interface for the user account.

All the features of the LinxConnect account can be modified using the Web-based LinxWeb interface. Currently, the interface is plain and to the point. There isn't any readily available help feature, and some settings could be better handled by the interface. But, relative to the phone-based menu system, the Web-based interface is a real time-saver when it comes to setting features. What would take a tedious hour on the phone, flipping through the menu system, is a simple point and click operation in LinxWeb. For example, you can rapidly manage faxes and voice mail, forward faxes to different numbers, and listen to voice messages. Also faxes and voice mails can be forwarded to another e-mail address, but, there is no auto-forwarding of messages to an e-mail address. Even though it is easy to manage your account and account settings using the Web, some things are still better done over a phone, such as setting greetings. Setting greetings over the Web requires a specific browser and PC configuration that some users may not have.

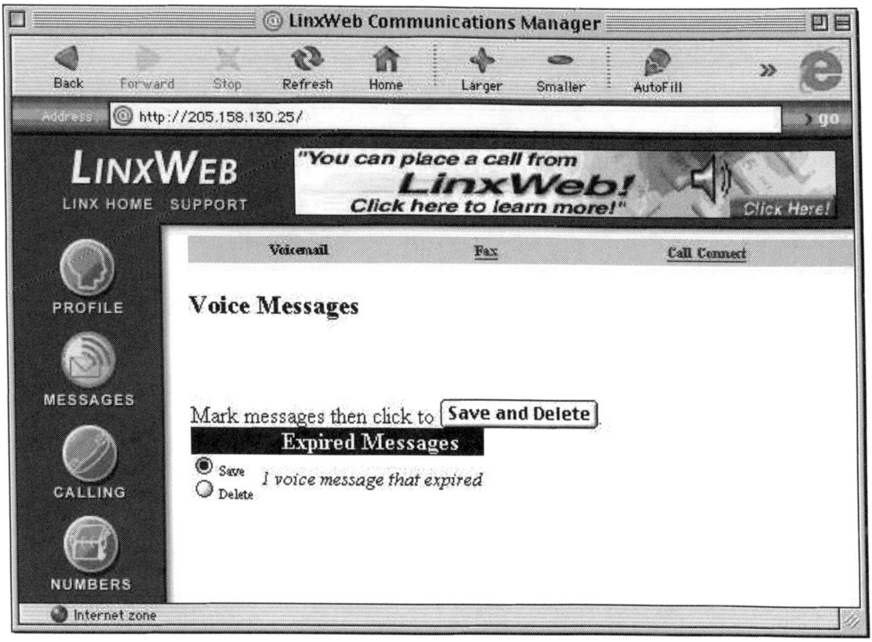

Figure 11-1: The LinxWeb interface can be used to manage and configure your account. Copyright Linx Communications, Inc.

11.1.2 Featues—LinxOffice

LinxOffice is a custom solution designed to meet the communication needs of offices of any size. It is a full-featured automated attendant front-end that can combine the features and functions of Linx's unified communications products, such as LinxConnect.

LinxOffice is targeted at any business with 2 or more employees. It is intended to provide, on a subscription basis, a virtual phone system that is complete and comprehensive.

LinxOffice presents an Automated Attendant to callers. As a front-end, it can route calls to any number. These numbers ideally would be a LinxConnect number to allow for fax reception and 'extension' specific call forwarding. LinxOffice also has PBX functions of its own and can integrate into existing Centrex or PBX systems. In addition to the call forwarding, other PBX functions allow users to set up conferences from any location or call into the system and use it to call out so as to take advantage of local toll rates. LinxOffice also includes a search by name directory and allows for extension dialing from within the system. In short, LinxOffice is an office-scale version of LinxConnect with Auto Attendant features. On its own it is indistinguishable from a regular PBX. With LinxConnect extensions, it is a powerful PBX solution.

11.2 Netergy Networks

The Netergy Networks Advanced Telephony System (ATS) is an IP-PBX solution designed specifically for service providers wishing to offer voice services to their customers.

The system itself runs on a cluster of Sun Microsystems Netra servers at a service provider's data center. At the customer's premise, a Netergy Media Hub is installed to interface the Netergy iPBX with traditional analog telephones and fax machines. Netergy also offers the following software applications that allow a customer to control and configure their iPBX:

- Netergy Communication Center—This application provides features that are used by most typical business users, including desktop call control, voice mail access, contact management, and more. The application runs in a standard Web browser so employees can access the system from the office, home, or a remote location.
- Netergy Switchboard—The Netergy Switchboard is specifically designed for receptionists or anyone who manages phone calls for a group of employees. The Switchboard makes it easy to see the status of phone extensions within the office and to transfer calls through a point-and-click interface.
- Netergy Administrator—The Administrator application allows Netergy iPBX customers to manage the configuration of their iPBX and handle all moves, adds, and changes. With a point-and-click interface, the Netergy administrator can setup new users, manage user profiles, and more.

The Netergy Networks ATS is one of the first hosted IP-PBX solutions to appear on the market. The convenience, scalability, affordability, and flexibility of a hosted solution guarantee that we will see more solutions like this in the future.

INSIDER PERSPECTIVE

The iPBX: Netergy Networks Hosted IP-PBX Solution

By Netergy Networks

To what extent does the iPBX hosting model appeal to end users, some of whom know little or nothing about VoIP systems and others who have heard the horror stories of low QoS (Quality of Service) and unreliable service?

In the following discussion, we will:

- Describe the target end-user market for iPBX hosting,
- Examine the existing telephony service options available to these customers, and,
- Illustrate the appeal of iPBX hosted voice services.

Target End-User Markets

The target market for iPBX hosting is high-density telephone customers with sites of up to 100 extensions. Enterprises of this size typically desire advanced telephony features and reliability, but don't want to incur the capital and personnel expenses necessary to install and manage a PBX system. These small- to medium-size enterprises include:

- Small- to medium-sized businesses
- Apartment buildings
- University/College Dormitories
- Local, state and federal governments
- Hotels and motels
- Convention centers and exhibition halls

In the near term, small- to medium-sized businesses are likely to account for the majority of iPBX hosting customers because of the combination of capital constraints and the willingness to adopt new technologies offering a competitive advantage. The primary need of these businesses is reliable voice and data communications, often around the clock. Beyond reliable service, these businesses also want voice services with the following characteristics:

- Low capital investment—Every dollar counts, with business owners preferring to leverage existing investments as much as possible and avoid large capital expenses for telecommunications equipment.
- Flexibility and scalability—Small- to medium-size businesses are more vulnerable to market cycles, often needing to grow or shrink personnel and expenses rapidly. These businesses would prefer a telecommunications solution which can flexibly scale with their needs so they are not stuck with too many lines, or too few.
- A total communication solution—Much has been made of the idea of combining all communication services into one bill. However, small- to medium-size businesses don't care about writing one check or two. They care about gaining a single point of contact for combined voice/data installation, maintenance and support issues. This single point of contact greatly reduces coordination costs and

the amount of time company personnel must spend solving problems, and increases loyalty to a single service provider.

- High functionality—Small- to medium-size businesses have size envy. They want to look and function like much larger companies in order to demonstrate the stability and trustworthiness that their customers demand. Using the latest technology is one way to have a "big" presence, and small- to medium-size businesses would like access to high-end PBX and CTI features for their messaging systems without the expense of purchasing and managing these systems in-house.
- Easy user control—Businesses want both the reliability of traditional, carrier-based telephone systems and the ability to easily re-configure the voice system and/or specific features to meet changing business needs.

Existing Voice Service Options vs. the Appeal of iPBX Hosting

Small- to medium-size businesses currently have several voice service options, including standard telephone service, Centrex service, and in-house PBX systems. Prior to iPBX hosting, the choice between these voice service options often forced customers to make difficult trade-offs between cost, reliability, and features versus scalability.

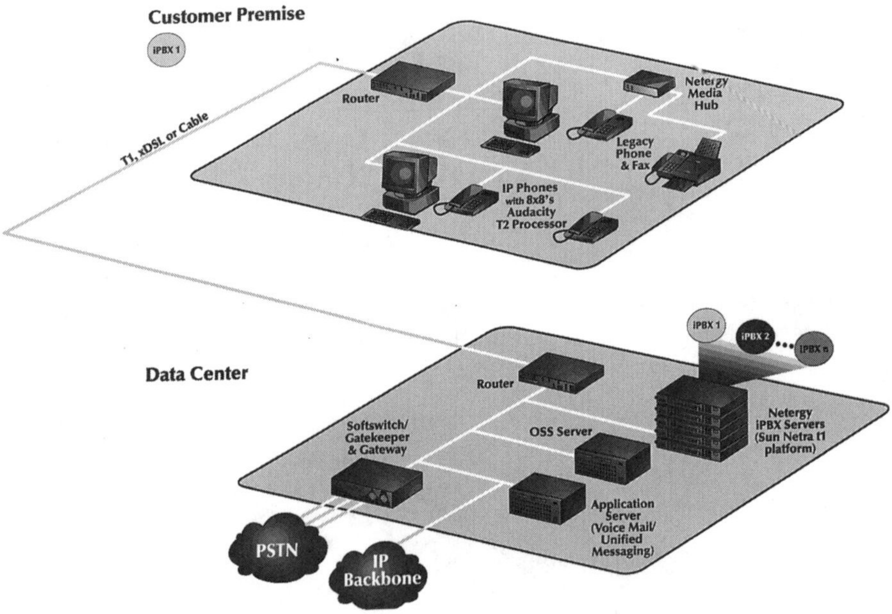

In a hosted configuration the iPBX is managed by a service provider.

A typical small business begins with standard telephone service from a local ILEC and may add Centrex services to gain basic call control features such as call-transfer, direct inward dialing, hunting or three-way calling with low initial capital investment. Centrex systems are a good way to outsource the complications of a customized telephone system, but Centrex features are difficult to use through standard telephone handsets. Businesses begin to examine alternatives to Centrex as they grow because costs for Centrex service scale linearly—each additional line costs the same.

As the number of extensions increases to 20 or more, businesses often consider investing in an in-house PBX system. A PBX system is a private telephone switch, located at the customer premise and connected to a service provider by external trunk lines. PBX systems provide a significantly expanded feature set over Centrex and the costs per extension for a PBX system decrease as new users are added. However, virtually all PBX systems in service today use a proprietary design and all major components, including phone sets, interface cards and control software, are also designed, manufactured and serviced by the same PBX vendor. Multi-vendor component choice is limited.

If a small- to medium-size business user charted their company's different telephony service options and costs as the enterprise grows, the cost comparison chart would look much like the one below:

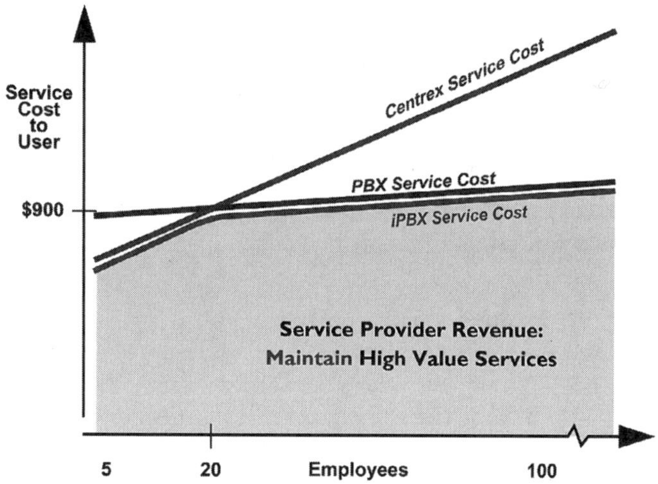

Cost comparison of Centrex Service, an in-house PBX, and a hosted iPBX solution.

The three lines represent the costs of the typical voice service options from which a small- to medium-size business owner would choose. The first option is installing traditional Centrex services and keeping Centrex as the company grows. The second line shows the costs for a PBX system. The iPBX hosting option, represented by the lowest cost line, is comparable in price to Centrex at low numbers of employees and is somewhat less expensive than even an in-house PBX system until the company grows.

From the perspective of the customer, the iPBX hosting model provides the lowest total cost for advanced telephony services as a business grows. With iPBX hosted voice services, combined with the convenience and reliability of Centrex and the flexibility and scalability of a PBX system, end-users are free to use their telephone system in the manner that makes the most sense for their day-to-day business rather than having to make the hard choices they currently face.

11.3 ThinkLink

ThinkLink, Inc. began in 1998 in response for the growing need for high quality, low cost Web-based integrated messaging. They first started with small beta testing of the service then it evolved into a nationwide beta test program. In the end of 1999, ThinkLink rolled out a final version of the service, which is different from the beta version in that it is lacking fax and e-mail capabilities. This is due, in part, because the whole service was revamped to accommodate a higher numbers of users. Also, the extensive beta testing may have given ThinkLink an indication of what part of its service customers used the most (voice mail and outbound calling). ThinkLink has plans to upgrade this final version to include fax and e-mail features.

What makes ThinkLink different from most of the other free Web-based integrated messaging services is that it incorporates both an inexpensive outbound calling plan and a call routing capability.

11.3.1 Target Market

The target market for ThinkLink's service is people who need a reliable integrated messaging solution that can be accessed while out of the office or home. Research from the Yankee Group revealed that most Americans have become harder to reach, except via voice or e-mail messages. Currently, in the US, there are over 1 billion voice messages a day. In a few years there will be 1.5 billion e-mail and 2 billion instant messages a day.

ThinkLink has been marketing its service to consumers directly and via a strong affiliate program. Their marketing strategy assumes that integrated messaging (voice, fax, and e-mail) is attractive to consumers, will self-market (grow virally) as people use the service, and will generate ongoing revenue, not via a subscription, but via outbound long-distance calling.

Their beta test period supported these assumptions. The beta version of the ThinkLink Web-based integrated messaging service, coupled with low-cost calling, generated high-volume use, sometimes higher than predicted. Furthermore, the majority of users visited their message center at least once a day, took advantage of the advanced features, and exceeded the national average of toll calls per month.

11.3.2 Features

This discussion is based on actual use. Users receive an "800" number, a local number, or both. Incoming calls can be automatically forwarded to single or multiple numbers (call forward or follow-me). Voicemails are encoded as either streaming RealAudio or WAV. The user can be notified of new messages via e-mail (e.g., to e-mail-capable devices) or paging. A page notification is considered a 1-minute outbound call. There is a well-designed address book for storing information, addresses, and telephone numbers.

Due to its very close relationship with ICG, a major long-distance Internet telephony company, ThinkLink users can place outbound calls from their accounts (for pages, forwarding, and original calls) to anywhere in the continental U.S. for 5 cents per minute. No other similar service we know of offers this price level on calls. Rates to other areas are competitive. Calling into the "800" number costs 10 cents per minute. Also, the numbers in the address book can be selected as speed dial numbers to help calling out directly from ThinkLink.

The main interaction with ThinkLink is Web-based. The interface is clean, responsive, and easy to get around. The user can also phone the account to manage settings, listen to voice mail, and make outbound calls. The quality of the call is indistinguishable from a regular toll call. The telephone- and Web-based interfaces are very good.

The beta version of the service also had a fax and e-mail messaging component. Faxes could be automatically saved or routed to another number. E-mail from POP accounts could be regularly collected. E-mail sent from the ThinkLink account could have a custom reply-to address (other than user@thinklink.com). Both e-mail and notification management could be customized via filters. ThinkLink says the current service will be upgraded to include e-mail and fax messaging— only they have been saying 'in the next two months' for the last six months.

There is one last feature that we couldn't test during the beta: Extensions. This feature allows users to link multiple ThinkLink accounts to a single number, similar to a voice mail system. We can see a small office using this feature to link different ThinkLink accounts to a local number and an "800" number to form a small voice mail system. And, combined with the call forwarding features, the Extensions feature can act as a virtual PBX. We do not know if the final version of ThinkLink will include this feature.

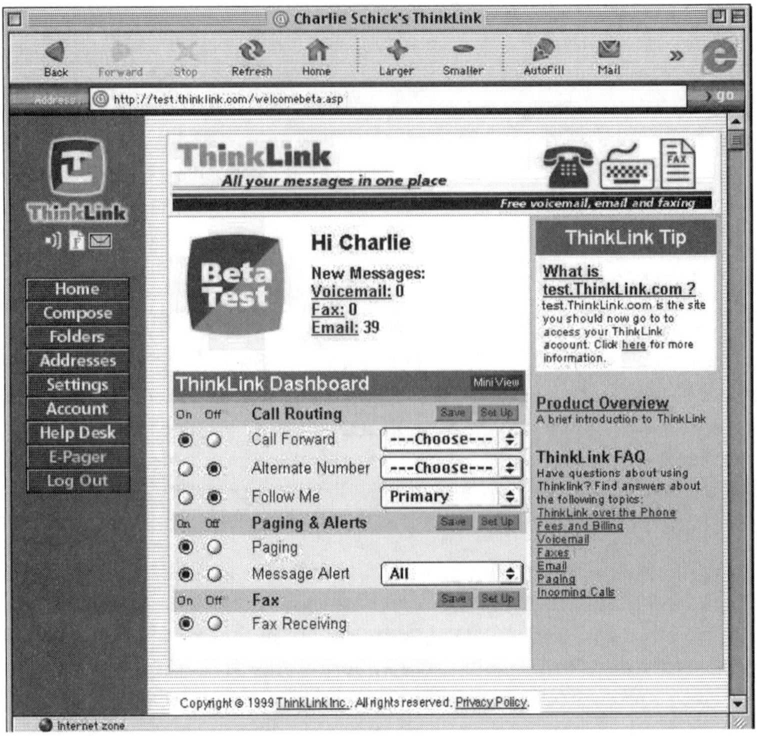

Figure 11-2: Setting ThinkLink account preferences via the web.
Copyright ThinkLink, Inc.

Figure 11-3: Accessing voice mail through the ThinkLink web interface.
Copyright ThinkLink, Inc.

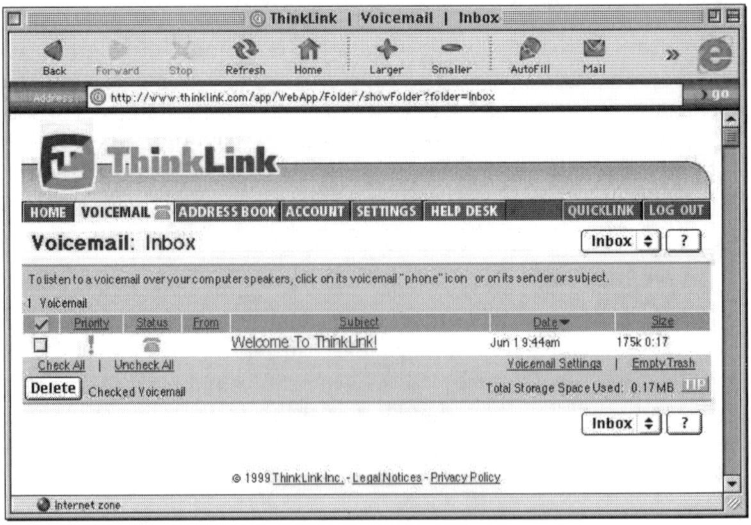

11.4 VirtualPBX.com

VirtualPBX.com is a complete outsourcing solution for companies that either do not want to manage their own on-site telephone system, or have many people scattered across several facilities. The only terminal equipment the user is required to have is a telephone (land-line or cellular). The service also provides an optional World Wide Web interface that can be used to configure and monitor the system.

The service is an especially attractive option for small businesses who want to avoid the up-front capital investment of buying a new phone system, and for companies with a highly mobile workforce. Traditional telephone systems, while they do allow users to forward calls offsite, generally do a poor job of handling the needs of a company where people are highly mobile, or work from multiple locations (e.g. telecommuters).

VirtualPBX.com presents a single system image to all outside callers. The company does not even need to have a primary office to use the service. The system tricks the callers into thinking they are calling a conventional office phone system, when in fact it is dispatching calls to people who are working from many far flung locations.

The service is also useful for companies that want to have a backup telephone service in case their office is damaged or their on-site telephone system has a catastrophic failure. In this scenario, the company normally would not use the VirtualPBX.com service, but instead would create a mirror image configuration in the VirtualPBX.com system (e.g. same extension numbers, same greetings, etc). Then, if they experience a major problem, they would simply have all of their incoming calls forwarded to the backup telephone number provided by VirtualPBX.com. The service would then reroute calls to users at their home phones, cellular phones or wherever they can be reached. Therefore companies that would otherwise have been unable to receive calls from customers could remain reachable, even in the event of the complete destruction of their primary office.

VirtualPBX.com emulates most PBX and voice mail services. A typical set up involves the company forwarding all calls to the VirtualPBX.com system and then the system handles all the calls from there. Once set up, all features are modified via a telephone interface. The company can set up:

- Multiple extensions
- Automated call distribution by list order, round robin, or priority
- A customizable auto-attendant

Each individual extension allows the user to control telephone contact information, such as cell phone, home, or work. The PBX will try these numbers to attempt to complete the call. If unsuccessful, the system will take a voice message for the extension. Each extension user can also participate in ACD workgroups (e.g. share incoming call workload with other users in the sales department). The system also has the ability to store incoming fax messages in each user's mailbox.

Other useful features of the VirtualPBX.com system include:

- Call screening, having the user announce his name
- Outbound dialing
- Mark an extension 'available' or 'unavailable'
- Dial-by-name to access the company directory
- Extension to extension dialing
- Customizable music on hold
- Direct inward dialing to individual extensions (employees) or workgroups
- Ability to transfer calls from remote location (i.e. an agent who receives a call on his cell phone can instruct the system to send the caller to another worker or workgroup)
- Integrated voice/fax messaging
- Web-based account management and configuration

Unlike many office phone systems, including some fairly expensive ones, VirtualPBX.com also offers ACD (automated call distribution) capability. This allows you to do things like grouping several users together and routing calls to them as a group. For example, suppose you have five sales reps. You could put these reps into a group called Sales (extension 1). Callers who select "sales" from the company's main menu will then be routed to the first available sales rep that has been idle for the longest period of time. This is a powerful tool for spreading incoming call workload across a group of people, and for increasing service levels by routing the caller promptly to a live person who can handle their request.

One of the best features of the VirtualPBX.com service is its very low initial cost. A company will typically spend less than $500 (often much less) to get completely set up with the service. The service is billed based on usage (per minute charges for incoming calls). This allows small businesses to test-drive the VirtualPBX.com service inexpensively, and to use it to prototype different ways of handling telephone calls. Even if a company decides that the service is not appropriate for their needs, the initial cost of the service is so low that it is insignificant compared to the cost of installing a conventional phone system. This makes it an excellent solution for companies that are unsure about their next system, or that are in transition between facilities. All in all, it's a high quality, low cost solution for companies that don't want to deal with the hassle and expense of managing their own phone system (or paying someone to manage it for them).

11.5 Webley Systems

Webley is a unified messaging and call management service, designed for personal or multi-user solutions. Users are provided with a personal '800' number for inbound calls and faxes. Users interact with the service via a voice activated or telephone keypad interface. There is a Web-based interface to upload contact lists and to manage messages and calling patterns.

Webley has the standard suite of features found in call management and unified messaging services, including:

- Rule based sequential (find-me) or simultaneous (blast-me) call forwarding
- E-mail can be managed over the phone and listened to via text-to-speech, forwarded to fax, replied to via voice attachment on an e-mail
- Call return after listening to voice mail
- Faxes can be read online, forwarded to another fax or e-mail
- Outbound dialing from system

Webley also has two features we have not seen in many similar products. Webley has true follow-me—Webley captures the caller ID number of a phone when told to 'Follow Me', facilitating the ad-hoc setting of follow-me numbers. And Webley prides itself in its voice dialing, allowing users to state an address book name or new number to initiate dialing.

11.6 Wildfire Communications

Wildfire Communications, Inc. was born in 1991 by Bill Warner, Rich Miner, Nick d'Arbeloff, and Tony Lovell. At the time, Bill Warner had purchased his first car phone and, as a busy executive starting up a company, he was spending a lot of time trying to get work done in the car. But, he found it was difficult to manage his calls from the car and even more difficult for people to reach him. Bill realized that he could create a system that would provide individuals the ability to manage calls from anywhere. Not one to reach for small goals, he and his colleagues decided to create a system that required sophisticated software using advanced speech-recognition technology.

Since then, Wildfire has received various awards, such as Product of the Year awards from *Internet Telephony* and *Teleconnect* magazines. In addition, *Teleconnect* honored Wildfire with its Editors' Choice 2000 Award. Wildfire is currently offered by

Pacific Bell Wireless in the United States, Orange in the United Kingdom and Bouygues Telecom in France. Wildfire is also in market trials with other operators throughout North America and Europe. Wildfire is available on an individual subscription basis from local Wildfire Authorized Service Providers.

Wildfire is targeted at mobile users with heavy telecommunications needs. Currently available for individuals, Wildfire is preparing an Enterprise product (late-2000). Presumably this will allow a closer interaction of Wildfire users, much like a virtual PBX.

11.6.1 Features

Wildfire's features include:

- Voice-controlled voicemail — Wildfire tags messages with a callback number via caller ID.
- Address book — Wildfire can store up to 150 contacts, with six phone numbers each. Contacts are accessed via the voice-controlled interface.
- Incoming call alert — When the user is using the system, either making a call through the system or listening to messages, Wildfire can notify the user by 'whispering'. The user can then decide whether to take the call or have Wildfire take a message.
- Voice dialing — In addition to dialing by name via the address book, users can dial numbers by voice.
- Call routing — Calls can be easily forwarded while connected to the system.

The service provider provides a Wildfire-enabled number. If the provider happens to be the user's wireless carrier, a new number will not be necessary. Also, all the functionality resides at the provider's servers, so any phone can be used to access account features.

Wildfire is the service that set the standard for the one-number call management service market. Its major differentiating feature is the speech-recognition interface. Since it is geared towards mobile professionals who most likely have cellular phones, a speech interface seems a better solution — a touch-tone interface would require the user to periodically listen to a IVR menu and look at the phone to touch a key.

The only caveat is that in our experience speech recognition only works in environments that are topically constrained. If there are too many options to choose from, especially if options are generated on the fly, then speech recognition fails. From a brief demo, it seems that Wildfire has a constrained and stable enough speech vocabulary to make the system more tractable.

Chapter 12

The Next Generation Telephone and Terminal Device

12.1 Introduction

Just as telephone systems are evolving with the introduction of new technology, so too are the terminal devices that sit on your desktop. Gone are the days when you'll have a dumb, overpriced analog telephone sitting on your desk. These new systems are designed to talk to Internet aware, multimedia communication devices that can blend voice, data and video.

Ironically, if you want to visualize where telephones are heading, it helps to go back to the future, to the 1960s. AT&T introduced the first videophone at the 1964 World's Fair in New York. Although the videophone was ahead of its time (and looked strangely similar to Apple's iMac design), it is a good example of what phones are likely to become in the near future.

The most important feature in next generation telephone systems will be the ability to talk to and transport calls to terminal devices (telephones, videophones, PCs, etc.) across a company's data network. This is important for several reasons, among them:

- The owner can mix and match different types of terminal devices (e.g. legacy analog, IP based voice phones, IP based videophones, PCs outfitted with NetMeeting, etc.).
- The physical media used to transport data from the phone switch to the end user is completely transparent to the phone system. This means the owner could mix and match different data transport technologies (e.g. fiber optics, high speed Ethernet, wireless Ethernet, and others) in their telecom network. The beauty of Internet protocols is they make the underlying technology invisible to the devices and software using them.

- The ability to make calls between very different types of devices. For example, a NetMeeting user could make a video call to a Polycom Viewstation, hang up, and then dial a voice only user on another extension. This interoperability between different types of devices allows the owner to outfit each user with the telecom services best matched to their needs.
- The ability to support high fidelity audio. Because these devices transport calls over high-speed data networks, they can be programmed to offer far higher quality audio (comparable with CD recordings).

When you consider all of these advantages, you can see why many manufacturers are migrating toward using the same technology that computers use to share data to transport voice and video phone calls. The result for you, as a buyer, is more options, better technology, and ultimately, lower costs.

12.2 The Next Generation Telephone

Telephones designed to work with next generation telephone systems come in the following basic flavors:

- Legacy analog telephones
- Digital telephones with dedicated telephone wiring (traditional PBX/Key phones)
- IP telephones which connect to a company's wired or wireless Ethernet data network
- Cordless telephones and site-based cellular phones

12.2.1 Analog Phones

Almost all of the next generation telephone systems work with conventional analog telephones (like the phone in your living room). These phones, while they do not have the sophisticated call management features found in digital phones, are easy to install, inexpensive, and readily available through retail and mail-order/Internet channels. They also come in many different form factors, so users can choose from a wide array of devices.

Analog phones can be loosely grouped into the following categories.

- Basic 1 or 2 line corded phones: these are basic desktop telephones with a fairly simple set of features. They are a good match for users who use the telephone occasionally, and don't require special features (such as the ability to juggle three or four callers at once).
- Full duplex speakerphones: these phones are designed specifically for high quality conference calls. The *Polycom Soundpoint* phone, an excellent example, is a 1 or

2 line corded telephone with built-in full duplex speakerphone. Unlike most low-cost speakerphones, the Polycom unit does not distort or clip conversations (producing the walkie-talkie effect common to low end speakerphone products).

- Single user cordless phones: these phones allow a cordless handset user to wander 50 to several hundred feet away from his or her base station (which is connected to the analog phone jack at the desktop).
- Multi-user cordless phones: multi-user cordless phone systems allow several cordless handsets to share a common base station, and to roam around within a larger facility without dropping calls. These behave like a poor man's cellular telephone network within the office.
- Cordless headset telephones: cordless headsets allow users to walk about the office while taking and placing calls and at the same time keeping their hands free. These devices are a good solution for people who are very mobile within the office, and who need to use their hands while they are talking (e.g. filing documents, handling boxes, etc.).

In terms of products, no particular product stands out for special mention. The line-up of products is also constantly changing. Hello Direct, Inc., a well-known direct marketer of telecom products and accessories, carries corded and cordless phones from many different vendors, and is a good source of information about the best products currently available.

12.2.2 Digital Telephones

Many of these systems also work with proprietary digital telephones. These digital phones are typically designed by the vendor, and only work with the vendor's system. They are also designed to offer extended features that are particularly useful to people who spend a lot of time on the phone, or need to be able to manage many calls at once. Some of the features offered by digital telephones include:

- Busy Lamp Field (BLF): this is a bank of LEDs which displays the status of outside phone lines and internal extensions. This enables users (e.g. the secretary) to see at a glance which lines are busy, and which employees are currently on the phone.
- Multiple Line Appearances: this feature enables a user to juggle many calls (sometimes as many as 8) simultaneously. Jumping from one call to another is a one button operation. This is a popular feature with people who are bombarded by multiple calls at once.
- Conference Calling: digital phones often make it easier for users to set up multi-party conference calls (though sometimes they make the process of setting up a conference call even more bewildering, it all depends on the vendor).

- Caller Identification: digital phones usually provide extended information about incoming callers, another popular feature with power users.

12.2.3 IP Telephones

IP telephones offer essentially the same features as conventional digital telephones. The main difference between the two is that IP telephones actually transmit phone calls over a company's data network (usually via a 10BaseT or wireless Ethernet connection). We'll discuss a few IP telephones later in this chapter.

These phones offer some important benefits compared to conventional digital telephones, among them:

- Remote sites: the telephones can be situated almost anywhere in the world provided there is a reliable TCP/IP data path between the phone system and the telephone. For example, this means a telecommuter with a DSL line could be set up to appear as though they were connected directly to the office phone system, even though they are off-site.
- Simplified wiring: conventional analog and digital telephones require a cable for every single telephone extension. IP phones can be multiplexed to share a single 10BaseT connection. For example, several IP phones could be connected to an inexpensive Ethernet hub which is, in turn, connected to the LAN backbone. This greatly simplifies wiring because one 10BaseT cable run can serve dozens of IP telephony clients if it is set up correctly.
- Independence from data transport technology: one of the great things about IP phones is they do not care how the data is transported from point A to point B, as long as the network is running the TCP/IP protocol. The data could be carried over fiber, 10/100BaseT, ATM, wireless Ethernet, whatever. This means the owner does not get boxed into using a specific networking technology as a prerequisite.

12.2.4 Cordless/Site-Based Cellular Phones

Specialized cordless and site based cellular phones allow a company to build, in effect, their own private cellular network inside the office. Some systems actually work with standard cellular phones (which can roam onto public cellular phone networks when users leave the office). Others use proprietary cordless telephones (some based on the DECT cordless phone standard developed in Europe).

These devices are rarely used across the board, but rather for specific users who are highly mobile inside and outside of the office.

12.3 Next Generation Telephone Market Overview

12.3.1 dba Telecom

dba Telecom makes a line of analog feature phones designed for use with next generation telephone systems (as well as Centrex service). The phone in Figure 12-1 is designed for use by a receptionist or other high traffic positions.

Figure 12-1: dba Telecom phones for next generation phone systems and Centrex service. Copyright (c) dba Telecom Corporation.

The phones provide most of the features found in high end PBX telephones (e.g. busy lamp field, call pickup, call transfer, etc). Although the telephones can be used on any analog extension, they work best in systems that have been modified to take advantage of the extended feature set provided by these phones.

12.3.2 Circa

Circa is a Canadian manufacturer of next generation telephone handsets. They make both analog and digital (IP based) telephones.

The analog phones, similar to the dba Telecom phones, provide extended features for power users, yet are still compatible with conventional analog phone systems and Centrex telephone service.

Figure 12-2: Circa makes analog and digital (IP based) phones. Copyright (c) Circa Communications, Ltd.

Their most interesting product is their line of IP phones that are designed to make and receive calls over a company's IP data network. They offer an extensive range of features (including the ability to use broadband connections to provide superior telephone audio quality). Among the key features of the IP phones are:

- Wide Band Capability (7 kHz) for superior voice quality
- Support for Megaco, H.323, and SIP VoIP protocols
- Display: 80 x 160 pixel graphic display

- Time and Date
- Call timer
- Message Waiting LED
- Dedicated feature keys
- Feature key mapping
- 4 Soft keys
- Scroll keys
- Context based volume controls
- Handset and Headset
- Full Duplex Speakerphone with Acoustic Echo Cancellation
- Echo cancellation
- Silence suppression and Comfort Noise Injection
- Jitter compensation
- Lost packet compensation
- G.711, G.723, and G.722 Codec support

12.3.3 Pingtel

The first iterations of PBXs were the proprietary systems we've since come to love to hate. Then, in a revolution led by vendors such as Artisoft and AltiGen, the logic of a PBX was placed onto a "open platform" PC. But, dumb phones were still connected to the system. What if the phones carried their own intelligence? What if you could do more than just program the extension number and set up a list of speed dial numbers? The xpressa from Pingtel is one such clever next generation phone.

Pingtel's xpressa

This is not a dumb phone. The xpressa was designed from the start to support PC-, Web-, and xpressa-based applications. No, this is not a Windows PC-based phone. This phone does not need a PC to work. The xpressa get its smarts from embedded Java. The phone control applications are written in Java and use JTAPI and Pingtel APIs. Pingtel offers a toolkit and a user interface style guide to help developers create applications for the phone. Alternately, the developer can use standard Java visual development tools. The phone is also compatible with MS-TAPI services and functions. It can communicate via IP with a PC for further integration, such as one-click dialing from a URL or from Outlook.

Figure 12-3: The Pingtel xpressa Java-based telephone. Copyright (c) Pingtel Corporation.

The phone includes a graphical interface (160x160 LCD) and an audio interface to help users with phone features. There are 11 programmable keys and many other buttons and knobs to scroll, access menus, perform standard call functions (call transfer, etc.), and a headset jack. Custom ring tones are actually .wav files. (Imagine the possibilities with that!) The phone also has a speakerphone and a separate message indicator light.

The Pingtel xpressa is interoperable with IP-PBXs, PC-PBXs, and other phone systems that provide IP-based voice to the desktop. Pingtel supports the SIP protocol and the next release of the phone will handle H.323 as well. The audio quality is toll quality. It uses the G.711 codec, silence suppression, white noise generation, and packet loss compensation.

Installation and management

The phone is plug and play. It can obtain an IP address via DHCP and search the network for the call controller, such as a SIP proxy or location server. A configuation file for the phone establishes the resident and dynamic applications and other phone settings. The phone does not need a central control unit; here the intelligence is at the edge of the network.

Administration can be performed directly from the phone, through the Pingtel deployment server, or remotely through a Web interface. This makes it easy for inventory control, upgrades, and application control. Indeed, Pingtel's Java-based deployment server is meant to help manage and maintain a large deployment of phones. Furthermore, with this software and the appropriate VoIP gateways, softswitches, and servers, a service provider could use these phones to offer Centrex-like service.

Finally, there are extra communications ports. There is an IrDA port to interface with computing devices, such as a Palm or laptop, and an extra Ethernet jack on the back allows users to connect a hub or computer, so a jack isn't lost due to the phone.

Why a Java phone?

A Java phone that can run applications opens up the possibilities for new voice services and Web-telephony integration. Java is a widely used development platform and Pingtel hopes that Java developers will create new services and applications for the xpressa. In addition to services, users will be able to personalize their phone by downloading and installing items on the phone and using them immediately. This opens, though, a whole other can of worms for the IS department.

Figure 12-4: Pingtel supports Java applications that run on a 160x160 display screen. Copyright (c) Pingtel Corporation.

Through programming, the phone can accomplish the obvious call control functions such as conference calls, transfers, hold, and follow-me. But there is much more. For example, a Pingtel phone could send contact information to Outlook, converting caller ID information into an electronic business card (vCard). Or, Outlook could manage a conference call via the phone, calling everyone from an Outlook-scheduled meeting. In addition to contact manager integration, a user could have the phone programmed to automatically time a phone conversation for billing purposes. Web-telephony integration could allow users to tie telephony into their site. For example, a user could download an applet that contains a visual version of a vendor's IVR system. This would speed interaction with a system. Or, via an applet, the phone could be part of a dynamic process, such as an auction.

Pingtel expects the IP telephony market to explode over the next few years. By providing the first Java-based VoIP phone, Pingtel is well-positioned to take advantage of this growth. It expects to begin shipping the xpressa in late 2000.

12.4 Software/PC Phones

12.4.1 Microsoft NetMeeting

If we were to select a single application that epitomizes the convergence of voice and data communication, Microsoft Windows NetMeeting would be at the top of our list. NetMeeting is a Windows application enabling voice, video, data, and application sharing over intranets and the Internet. Recently, Microsoft released NetMeeting 3 with a streamlined user interface, improved voice and video performance, Internet telephony gateway support, more comprehensive support for application and desktop sharing, and a range of other features destined to make this version of NetMeeting even more popular for corporate applications, call centers, and general purpose collaboration.

Ever since Microsoft announced the first version of NetMeeting in mid-1996, the focus of NetMeeting has been on real-time multi-party collaboration and application sharing. In characteristic fashion, Microsoft never promoted NetMeeting as a voice or video over IP application. Rather than focus on technology, Microsoft marketed NetMeeting as a standards-based solution for enhancing interaction and communication between remote users. With NetMeeting 3, Microsoft has taken a significant step toward making multi-media conferencing a ubiquitous application.

With NetMeeting 3 almost every feature has been rethought. Even the name has changed to Microsoft Windows NetMeeting to reflect the fact that NetMeeting is now bundled with the Windows operating system. Microsoft Windows NetMeeting is a 100% software application (video capture hardware is optional) designed for desktop users. To learn about the changes and improvements in NetMeeting 3 and to discover if NetMeeting may be the right platform for your next generation call center (or for just calling home to Mom), read on.

Keep it Simple

Anyone who is familiar with previous versions of NetMeeting will be pleasantly surprised by the simplified user interface of this new version. On the surface, NetMeeting 3 is a radical departure from previous versions.

For starters, the application appears in a much smaller window and uses a simplified set of graphical control buttons. The interface is clearly oriented around the concept of a "meeting" or "conference." With the default user interface configuration, the top-half of the NetMeeting window displays an incoming video image and the bottom portion of the window displays a list of all conference participants.

If you prefer a smaller window, the NetMeeting interface can be streamlined to display only the video image or the conference participant information. Additionally, with the new "picture-in-picture" video feature you can preview your own image in the corner of the image you're receiving (or you can chose to display your own image in a separate window). The incoming video image can also be resized to an almost full-screen size.

Auxiliary functions, such as searching a directory or viewing call history, now have their own dialog boxes. Gone are the days of toggling a single screen between the current call, a directory listing, speed dial information, and call history.

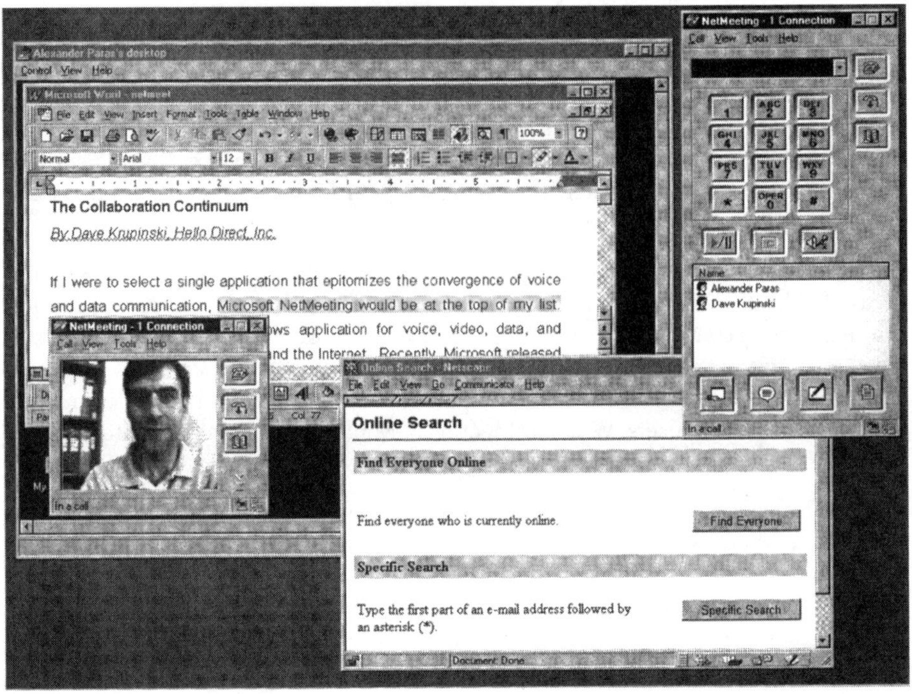

Figure 12-5: An application sharing and video conferencing session with Microsoft Windows NetMeeting.

As well as redesigning the user interface, Microsoft also simplified the use of directory servers. Instead of logging into a directory and waiting while the entire directory listing downloads to your computer (often a tedious activity), Microsoft has added support for HTML-based directories, which are searchable via the Web. A Web-based form for searching or browsing a directory can be accessed either from a NetMeeting dialog box or directly through a Web browser. (Features such as this are one of the reasons that NetMeeting 3 now requires Microsoft Internet Explorer 4.1 or greater to be installed on your computer).

Share and Share Alike

NetMeeting 3 also incorporates greatly improved support for application sharing. Version 3 provides a more powerful control panel for managing the applications that you wish to share with other users. Additionally, applications that are running on someone else's computer will now appear in a separate window on your desktop. This makes shared applications much easier to manage and to distinguish from your own applications.

Another significant new feature in NetMeeting 3 is desktop sharing. With desktop sharing you can allow another conference participant to take complete control over your computer—or you can take control over someone else's computer. Of course, you'll want to use this feature with caution and only share your desktop with someone you trust. (When sharing your desktop, you can take control away from the remote party at any time by pressing the ESC key).

NetMeeting 3 also incorporates a powerful variation of desktop sharing called "Remote Desktop Sharing." This feature is similar to the support provided by remote control software programs, such as pcAnywhere. Remote Desktop Sharing is great for applications such as controlling your office computer from home or for remote administration of an offsite server. When Remote Desktop Sharing is enabled, NetMeeting runs in the background and will automatically respond to incoming sharing requests. Of course, this feature is password protected and only runs over a secure NetMeeting call. (Security has also been improved in NetMeeting 3).

Application and desktop sharing are not only important for multi-party collaboration, but they also have tremendous potential for call center applications, help desks, technical support services, and training. Just imagine speaking with a tech support representative, while she demonstrates how to configure a new application directly on your computer. When combined with an IP-based call distribution application, NetMeeting looks more and more like it will evolve into a standard desktop component for the next generation call center.

Additionally, the NetMeeting 3 software development kit (SDK) will allow developers to directly integrate NetMeeting functionality into a Web page for more interactive customer support applications.

NetMeeting 3 also provides support for whiteboarding, transferring files, and multiparty chat.

H.323 Gateway Support

The H.323 gateway support of NetMeeting 3 is designed primarily for use with private or corporate Internet telephony gateways.

A Step Further

There are a number of third-party applications that enhance the functionality and work around some of the limitations of NetMeeting 3. Multi-party conferencing with NetMeeting still requires a third-party conferencing server, such as White Pine Software's MeetingPoint server. (If you'd like to try multi-party conferencing, you can view a list of public MeetingPoint servers on-line at CU-SeeMe World at www.cuseemeworld.com). One feature that has been removed from version 3, is the ability to switch the incoming video image between the multiple participants of a conference. With version 3, you can only exchange video with one other conference participant.

Another shortcoming of NetMeeting is that you need to have NetMeeting running to receive an incoming call request. Although you may be online, if NetMeeting is not running, another user will not be able to call you. This is where applications such as AOL's ICQ play an important role. ICQ allows you to view the online status of other users and exchange real-time messages. If you want to call someone with NetMeeting, just send him or her a quick note with ICQ asking them to make sure NetMeeting is started. You can then initiate a NetMeeting call directly through ICQ.

The more we use Microsoft NetMeeting 3, the more we understand the power and potential of this application for both professional business applications and personal use. Microsoft has always set ambitious goals for NetMeeting and with NetMeeting 3, they've taken a giant step toward reaching those goals.

12.5 Videoconferencing

Videoconferencing, once found only in the boardrooms of America's biggest companies, is now a mainstream technology. In the past year, videoconferencing systems have broken the $2000 price barrier, and by the time you read this, professional grade videoconferencing systems will be on the market for under $600.

Since the technology has been around in some form since the 1970s, why is videoconferencing just now beginning to take off? There are several reasons for this. The two most important factors driving this are 1) the dramatic improvements in computing technology in the past ten years, and 2) the development of technical standards that allow systems from different vendors to talk to each other across both ISDN and IP data networks. These standards have brought about competition, which has really accelerated the development of high quality videoconferencing products.

Since most of these systems conform to the standards for routing voice and video calls over IP networks, they can also be used as terminal devices with many next generation telephone systems. Especially interesting are the desktop videophones (such as Motion Media's videophone) that will combine videoconferencing and desktop telephony into a single user-friendly device.

12.5.1 Polycom Viewstation

One of the leaders in the videoconferencing arena is Polycom. Their Viewstation family of videoconferencing products provides a full range of solutions, from entry-level videoconferencing systems for the desktop to high-end systems for use in a boardroom.

Figure 12-6: The Polycom Viewstation videoconferencing unit. Copyright (c) Polycom, Inc.

The Viewstation systems support the H.323 standard for making video calls over an IP network, and can also make and receive video calls over ISDN telephone lines. They can be used to place both low speed, fair quality calls (128Kbps data rate per call), as well as near broadcast quality calls (at 384 to 768 Kbps data rate).

The systems are fairly turnkey and connect to a conventional TV set or monitor.

Polycom will also be releasing a desktop videoconferencing product, ViaVideo, in Fall 2000. This unit will connect to a desktop PC via a USB (universal serial bus) cable, and will offer professional quality (30 frames per second) video. Best of all, it will be priced at a recommended retail price of about $600 making it, by far, one of the videoconferencing best deals on the market.

12.5.2 C-Phone

C-Phone Corporation, like Polycom, focuses on set-top videoconferencing systems for use in conference rooms. Their key products include a set-top videoconferencing system that connects to an ISDN phone line.

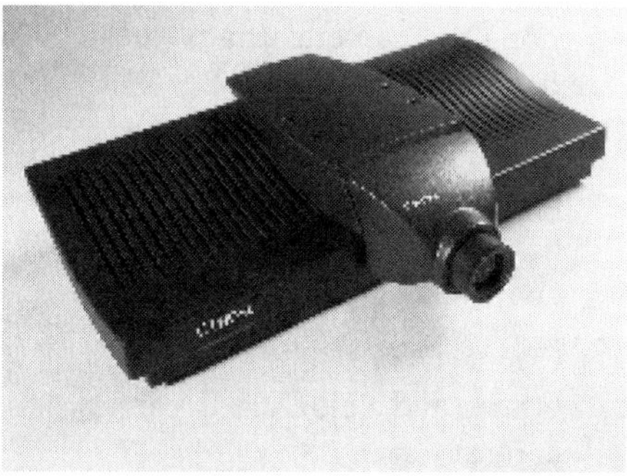

Figure 12-7: C-Phone's set top videoconferencing unit. Copyright (c) C-Phone Corporation.

The units support industry standards for videoconferencing over ISDN phone lines, so these units can make and receive calls from other standards-compliant videoconferencing systems, such as Polycom's Viewstation, and PC conferencing kits such as Intel ProShare.

12.5.3 Motion Media

Motion Media is a UK based company that makes a family of ISDN based videoconferencing products. Their most interesting product is the mm225 videophone (see Figure 12-8). This ISDN videophone is compact enough to sit on your desk and enables the user to make and receive ISDN voice and video phone calls.

Figure 12-8: Motion Media ISDN Videophone. Copyright (c) Motion Media Technology, Ltd.

The units price out around $2000, and are completely turnkey. No additional hardware besides the phone and an ISDN line to connect to are required.

12.5.4 Videoconferencing with NetMeeting

Microsoft Windows NetMeeting can also be used as a videoconferencing system. NetMeeting will accept video input from most Windows compatible video cameras (including new, inexpensive USB videocameras). This makes adding video to NetMeeting voice and data sessions a snap. Just connect a compatible camera to your PC, and you'll be able to send video when you set up a NetMeeting phone call. If both parties have video cameras hooked up, NetMeeting will establish a two-way video call.

This is a great solution for people who want to place video calls across their company data network, without having to go to a specially equipped conference room. NetMeeting works with many low-cost PC cameras, so the cost of adding video to a typical PC is less than $150.

Among the cameras that work well with NetMeeting are (this is not an exhaustive list):

- 3Com Home Connect: this USB video camera from 3Com is easy to install, and provides good quality video. It is not broadcast quality, but it is generally more than adequate for desktop videoconferencing. With a price tag of under $200, it's an inexpensive solution as well.
- Connectix Video Cameras: Connectix makes a line of inexpensive black and white, and color video cameras for use with PC and Mac videoconferencing applications.
- Kodak Digital Video Cameras: Kodak makes a line of digital still and video cameras, some of which can be used with a PC. The Kodak DVC323 is designed specifically for use with NetMeeting and related applications. This camera supports the USB interface.
- Polycom ViaVideo: due out in Fall 2000, ViaVideo is a complete audio/video conferencing kit that includes a USB camera and full duplex speakerphone. Priced at around $600, the kit enables PC users to participate in high quality video calls via their LAN or the Internet.

While videoconferencing systems have traditionally been completely separate from most phone systems, the trend today is to integrate these technologies into a company's communication infrastructure. The ability to make and receive calls over a LAN, using the H.323 standard, makes it possible to connect video terminals to next generation phone systems (which will then treat them like other terminal devices). This means that a user on an analog, voice-only phone will be able to dial another user who is using an IP based videophone. The phone system will act as a go between, and will enable people who have widely differing desktop equipment to talk to each other.

12.6 Summary

So, what type of terminal device is the best to use? It really depends on each individual user. The nice thing about the proliferation of options brought about by next generation systems is the ability to mix and match different types of desktop devices. Instead of being forced into a "one size fits all" system based on proprietary hardware, users can cobble together an integrated, multimedia communication system using components from many vendors.

Chapter 13

Next Generation Approaches of Traditional PBX Manufacturers

Traditional PBX manufacturers are not sitting still while next generation PBX vendors try to overtake them. The leading traditional PBX vendors know how to design excellent voice communication solutions. And, while they are hesitant, they know that they must move into next generation (especially IP-based) PBXs. We've rounded up the usual suspects and found all of them providing some sort of next generation product.

13.1 Alcatel

Alcatel's IP-based OmniPCX 4400 is an enterprise communications system. It has been designed with groups and individuals in mind and is complemented by PC-based telephony applications. The OmniPCX works over the Internet, private IP networks, ATM, or public ISDN voice networks. It supports and can integrate landline with wireless technologies, such as DECT and public cellular networks, for mobile applications.

The phone system has standard PBX features based on telephone prompts, such as voice mail, directory applications, automated attendant, and messaging. More sophisticated features, such as ACD and IVR for call centers can be integrated into an enterprise's IT platform and software in a client/server model. Other PCXs can be linked to the OmniPCX over any type of network, communicating via Alcatel's proprietary Alcatel Business Communication (ABC) protocol. ABC is a fail-safe protocol that can provide advanced user features across a network of PCXs.

The OmniPCX 4400 allows network management via a Windows interface or a Web browser. The standard network security features, such as trusted host and TCP wrappers, are available and the system can notify the network manager of illegal access attempts. Also, the OmniPCX is compatible with third-party firewalls. Other network features include simplified IP address management, directory synchronization across the network, and billing and usage reporting via e-mail.

13.2 Comdial Corporation

One of Comdial's next generation solutions is the FXS—a Windows NT-based PBX bundled with a dual-processor server. The FXS is a full-featured PBX that can support from 16 to 400 users with a specialized switching array that gives all ports simultaneous access through the system (from a single cabinet with 112 ports to a three-cabinet system with up to 544 ports). The system software comes with more than 200 built-in telephony features, such as multi-party conferencing and automatic route selection, and is modular, allowing for growth. Other applications, such as voice mail, unified messaging, automated attendant, and IP telephony, come with the server, and can be activated as needed. The system can also work with cordless and wireless solutions, such as the Scout 900MHz wireless telephone, pagers, and cellular phone systems.

Comdial also offers some software solutions that connect PBXs via an IP network. The first one, Comdial's System Networking connects up to 11 Comdial Business Systems into a single communications platform. This allows users to centralize applications, such as voice mail and IVR. Furthermore, these connected systems route calls over IP, saving a business toll costs. The second software solution, called CT Voice, provides for fax-to-fax, telephone-to-telephone, and computer-to-telephone communications. It provides for intelligent routing, carrying a call across the IP network to the CT Voice system nearest the destination and completing the call over the PSTN or local PBX. The software also includes a complete suite of software to support routing, billing, accounting, and network management.

13.3 Lucent Technologies

Lucent Technologies recently announced that it would spin off its PBX divisions. Most of its efforts in next generation phone systems have been in switches and network devices. Prior to the spin off announcement, Lucent had announced a number of products that would add IP capabilities to its existing PBXs. One of the announced products is the IP ExchangeComm (IPEC). The NT-based IPEC is designed for businesses with up to 96 extensions and as many as 200 stations. The major components of the IPEC are:

- the IP Exchange CallManager, which provides call control over the LAN/WAN.

- the IP ExchangeAdministrator, a Web-based graphical interface for secure access and network management.
- the IP Exchange Messaging, for enhancing the built-in voice messaging and automated attendant applications.
- the IP Exchange TelConnect, to connect calls to and from the PSTN via an IP telephony gateway or server.
- and the IP ExchangeAdapter, a stackable, eight-port unit to integrate the system with other Lucent solutions.

IPEC has an open architecture and APIs are available with a Software Developers Toolkit for the design of applications.

The IPEC system has all the standard PBX features, such as conference calling, call forwarding, automated attendant, and voicemail. In addition, the system can access directories via LDAP, store CDRs for analysis and reporting, and route calls via the PSTN or via IP.

13.4 Mitel Corporation

Mitel's next-generation PBX is the NT-based Ipera 2000. Mitel claims this IP telephony solution brings 99.999 percent reliable call control to the Windows NT environment. The Ipera 2000 supports IP and digital telephones, analog and digital trunking, and seamlessly networks with other Mitel phone systems (and any QSIG-compatible PBX). The system is designed for SMEs with up to 100 people. Users have desktop access to over 500 system features, such as automated attendant, skills-based ACD, unified messaging (accessible via Web or telephone), and cost control and management. The Ipera 2000 can also integrate with third party NT-based applications, such as voice mail and call accounting. Administrators can use Web-based management tools that use LDAP, simplifying moves, adds and changes, directory management, alarms management, database back-ups, remote maintenance, and remote programming.

Two new Mitel IP phones—the Superset 4015 IP and the Superset 4025 IP—have access to all the system features. Two other similar phones, the Superset 4125 and 4150 connect to PCs using RS-232 connectors. These phones can use TAPI-compliant computer telephony applications, such as Personal Information Managers and custom applications, to add further functionality to the PBX.

13.5 Nortel Networks

Nortel is no stranger to next generation phone systems. In addition to providing IP-voice-savvy switches and CTI toolkits for its Meridian phone systems, Nortel provides the Business Communications Manager (BCM). The BCM is an IP-enabled integrated voice and data communications system for small and medium businesses. It is compatible with packet and circuit switching, integrating PBX functions with VoIP functions in a modular system. BCM also provides for various services, such as, networking and QoS. It comes as a hardware/software package and runs on Embedded Windows NT, but voice service call processing is handled independently of Window NT.

Many applications come pre-installed on the BCM and can be rapidly activated with a key code. The free key code allows you to test the software before purchasing the component. The components provide features such as unified messaging, automated attendant, call center functions, CTI, message networking, and attendant console.

13.6 Siemens

Siemens has not announced an IP-PBX, per se. They have developed products that integrate with their Hicom line of PBXs. For example, the Hicom Xpress is an IP-based migration scheme to integrate PBX-based communications into an existing IP network. The Hicom Xpress products consist of a Telephony Internet Server that acts as an IP Gateway, and the Hicom Xpress @ LAN to integrate the Hicom 150E communication servers with a LAN.

The HiNet line of products adds functionality to the Hicom Xpress IP-based voice network. There is the LP 5100, an H.232 IP telephone; and the TA 1100, an analog IP adaptor (for faxes and modems). The closest Siemens comes to offering an IP-based PBX is the HiNet RC 3000.

The HiNet RC 3000 is an IP PBX client-based solution. It is intended to be used in conjunction with a standard PBX, either for migration to IP-telephony or to provide PBX access to remote offices or workers over IP. It is also intended to provide IP-based functions to PBX calls, such as Web-based collaboration with a customer calling into a help desk. Siemens does not intend to have the RC 3000 act as an independent PBX. The RC 3000 currently does not have certain standard PBX features, such as automated attendant, DID, or voicemail.

The RC 3000 is one of many products Siemens is developing around IP. They do not see traditional, circuit-switched PBXs becoming obsolete. But, they do see the necessity of integrating IP and ATM functionality into traditional PBXs. Siemens intends to evolve its Hicom and HiNet products into a wholly integrated managed network.

13.7 Summary

Understandably, many of the products provided by traditional PBX vendors are meant to integrate the vendor's legacy PBX with IP networks. Considering the caution usually displayed by large and entrenched corporations, this isn't surprising. These integration tools will play an important role in migration to fully IP-based and PC-based system architectures. As these tools are deployed, the vendors will be able to learn what customers want and what works best. There is a reason these corporations dominate the PBX market. While it seems like the nimble next generation system manufacturers may be overtaking the legacy vendors in the next generation communication system market space, we believe that as the traditional vendors learn more, the little vendors will be bought outright. Let's see how the market will fare then.

Chapter 14

A Look Into the Future

In this chapter we present a summary of our thoughts on how PBXs and phone systems might evolve over the next 10 to 20 years. Based on our view of where the industry is heading, we provide some tips and advice to help you make a more informed decision about selecting a next generation phone system today.

14.1 Dave Krupinski's Outlook

The phone system industry is currently in an exciting period of growth and change. At the same time, it can be an intimidating period for IT managers, small business owners, and other potential phone system buyers. It seems almost impossible to predict which solutions, architectures, and manufacturers will become the *de facto* standards of tomorrow. Yet, the selection of a phone system is a strategic, long-term investment. Is it wiser to select a "safe" traditional PBX solution? Or, does it make more sense to "risk" selecting a next generation solution overflowing with promises and potential. By looking at the lessons of the past and the technology trends of the future, hopefully I can offer some guidance.

14.1.1 Learning from the Past

In many ways, the state of the phone system industry today is very similar to the period in the mid- to late-1980s, when computing applications based on the personal computer were beginning to prove that they could replace mainframe systems and software. At that time, every IT manager saw the potential of the new PC-based client/server solutions, but also understood the risks of running strategic, mission-critical systems on an unproven architecture. Today, PC-based business applications have completely replaced mainframe systems in most organizations. In a similar way, I believe that next generation phone systems are poised to replace the traditional mainframe-like PBX.

However, that's not the only lesson we should take away from looking at the past. It's also instructive to take a close look at how organizations introduce PC-based client/server applications. As an IT consultant during that period, here are some of the lessons I learned and how these lessons might be applied to the evolution of the voice communications infrastructure:

- Evolutionary, not revolutionary—The transition from mainframe applications to PC-based systems didn't occur overnight. It was a gradual process of pilot projects, trials, learning, confidence building, and, ultimately, success. In the same way, an organization shouldn't completely replace their traditional PBX platform all at once. Rather, an organization should identify pilot projects and begin trials of next generation voice communication solutions internally or on a departmental level. Only then can an organization make an informed decision based on first-hand experience and learning. An evolutionary approach, which is aggressive about embracing new technology, but conservative in its deployment, if often the best approach.
- Consider integration strategies—Just as PC-based computing applications need to co-exist with legacy mainframe systems, so do next generation voice communication solutions need to integrate with traditional PBXs. Explore and test integration strategies. Many next generation solutions are designed to operate alongside or "behind" a traditional PBX. Introducing a next generation phone system is not an all or nothing proposition. Look for ways to introduce next generation solutions one step at a time.
- Build the right team—An IT manager is only as good as his or her team. Consider the skills required to evaluate, implement, and support next generation solutions. Make sure your team is ready to meet the challenge. An inter-departmental taskforce is often a practical way to raise awareness of next generation voice communication solutions and verify that the solutions under consideration will also meet the business requirements of the users. Next generation voice communication solutions are more than just plumbing—they also provide support for enhanced applications.
- Start today—The transition to next generation voice communication solutions is real. Don't wait until an urgent need arises before you begin testing and evaluating next generation solutions. Take the initiative and plan ahead. Evaluate multiple solutions and understand their features and limitations. Then, when the need for a production system arises, you'll be confident and prepared.

14.1.2 An Eye toward the Future

Which next generation architectures and solutions will be most prevalent 10 to 20 years from now? No one's characteristics and how I see the phone system industry evolving in the years ahead: crystal ball is perfect, but I do believe there are specific characteristics and features that make some solutions more future-proof than others. Below is a summary of these

- Standards-based modularity. Systems based on a modular, plug-and-play design are destined to have more longevity than monolithic, all-in-one solutions. Voice communication solutions that separate core functionality into modular units integrated via standard protocols offer the greatest potential of dominating the market. A standards-based infrastructure will ultimately guarantee the success of next generation solutions by creating a vibrant and dynamic marketplace with a wealth of options for the consumer.
- Specialization. In a world of standards-based interoperability, multi-vendor solutions will prevail. Vendors will begin to specialize in the area where they have a competitive advantage. Customers will select the components that meet their specific requirements.

  Eventually, there will be a separation of the "infrastructure" from the "application" (with the possibility of additional layers of separation in between). Some companies will specialize in the hardware and software to take real-time communication from point A to point B within an organization, while other companies will build the applications that provide enhanced control of the media stream. In the years leading up to industry-wide interoperability, look for increased partnerships between the leading "infrastructure" (i.e. the IP-PBX vendors) and the leading "applications" vendors (i.e. the PC-PBX vendors).
- Distributed architecture. A distributed architecture can provide scalability, administrative flexibility, enhanced reliability, and transparent feature access. Consider the Ericsson WebSwitch. Each WebSwitch unit can be distributed so that it is physically near the users it serves. At the same time, the system administration can be performed from anywhere on the LAN. Since there is no single point of failure, the entire solution is more resilient.

  In addition, a distributed architecture means having complete access to all the services of a solution from anywhere. Telecommuters, remote workers, and road-warriors with handheld phones should have the same access to features as on-site users.
- Multi-media support. Throughout most of this book, we have focused on real-time *voice* communication. Yet, the distinction between real-time voice, video, and data communication is blurred. On an IP infrastructure, there are really just two types of communication: real-time communication and everything else (even this distinction is arguable in a world of abundant bandwidth). Accordingly, next generation communication solutions will increasingly handle all real-time communication—voice, video, or data.

In 20 years time, the phone system as we know it today will no longer exist. Real-time communications will be the feature of the packet-switch data network infrastructure. This infrastructure will be dotted with advanced application servers providing all of the advanced communication applications that we use today—unified messaging, ACD, auto-attendant, IVR, video-conferencing, follow-me roaming, whiteboarding, etc.—and others that will only become apparent as this next generation infrastructure develops.

14.2 Charlie Schick's Predictions

The evolution of the private and public telephone systems over the last few decades does not really offer any guidance in prognosticating the next few decades. Why? In the last year or two technological and commercial transformations in the telecom and datacom industries have created a truly converged telephony industry, unlike anything in the previous 90 years. The rules by which voice communication solutions will evolve have been re-written completely. At the heart of the new evolutionary process is IP and packet-switched networking.

14.2.1 Voice over IP

In the past year, Voice over IP (VoIP) has evolved from a technology for advanced users to a core technology for the largest telcos. Indeed, all the large telcos and major traffic carriers have shifted towards an all IP network.

(Easy) Prediction One: The cost savings and flexibility of an all IP voice system will be the foundation for the next generation of telephony.

And, since voice will still dominate data in traffic volume, there will still be a need for some sort of IP-enabled communication server or "branch exchanges."

14.2.2 The IP-PBX

Private branch exchanges have evolved from using a human (switchboard), to a specialized computer (PBX), to an off-the-shelf PC (PC-PBX). As the VoIP network filters down from the long-haul carriers and into some form of local exchange, there will be a need for an IP-enabled PBX—the IP-PBX. The IP-PBX is already here. Various companies, such as Nokia and Netergy Networks, are trying to sell the idea (and products) to ISPs. ISPs have a large base of IP-connected customers. The IP-PBX or netCentrex allows an ISP to offer telephony services (Centrex-like) through the same pipe as the data services. Indeed, Nokia not only sells the netCentrex servers, but also sells the IP-based business phones. Companies like Cisco make and use IP-PBXs (really just large routers with IP-telephony call routing software) to provide telephony service for the whole global organization, with ensuing cost savings.

Prediction Two A: The IP-PBX is here and over the next 5 years will become the prevalent, if not the dominant method, for businesses to manage telephony services. ISPs, cable companies, and telcos will provide telephony services.

But, there's more. As less expensive high-speed data access in the form of cable modems and xDSL becomes more prevalent, data service providers will offer telephony over the same pipe.

Prediction Two B: Ten years from now, a single pipe will offer commercial and consumer telephony. New users will have one data pipe versus three (telco, cable, and data).

With the cost of providing data pipes falling, per minute rates for voice (or data) traffic will also plummet. No longer will voice/data traffic minutes be a source of revenue. How will the telcos make money in this VoIP world?

14.2.3 Service Providers

There are more messaging modalities than just voice. Indeed, there are communications servers, large and small, that integrates e-mail, voice mail, and faxes into a Unified Messaging architecture. But, these still utilize the public telephone network to deal with voice and are not capable of dealing with other communication means that exist on the 'Net, such as Instant Messaging. Furthermore, to provide Unified Messaging services, various different types of devices (fax, voice mail, e-mail, and Web servers) are needed. Future communications servers in an IP network will need to be able to utilize all forms of communications information, such as, user location, connection methods, message sending and retrieval, and user privacy.

Prediction Three: Communication servers will be at the heart of ALL communications and the global IP network will provide the connections.

14.2.4 Provided Services

With IP-based communication servers, network operators can provide enhanced services over the global IP network. Indeed, the mad dash to IP-enable everything from refrigerators to cars is really an attempt to extend the IP-network and increase the opportunities to offer enhanced services. We are already seeing e-mail, Unified Messaging, paging, back-office operations, and many other in-house services being outsourced and purchased over the Internet. With an IP-based communications server the PBX, call forwarding, call waiting, and other regular late 20th Century call features can be outsourced over the Web. Furthermore, these services can be integrated and served over the local LAN.

Prediction Four: IP-based communication servers will enhance, improve, and tie together the deployment of IP network-based enhanced services.

Is that all? No, the communication servers of the next few decades will also know quite a bit about our whereabouts. These communication servers will also manage our virtual presence.

14.2.5 Instant Messaging

Instant Messaging (IM), better known through AOL and ICQ, provides instant exchange of messages between subscribers. The heart of the IM interaction is the Presence Manager (PM). The IM client manages the settings of the PM server—setting location, contact information, privacy, and presence. But, current implementations of IM are restricted to PC-PC communications. AOL and others have announced plans to extend IM to wireless devices, such as PDAs and cell phones. Ericsson is testing a carrier based PM server for WAP-phones, called iPulse. Over the next few years IM will broaden to include all forms of communication devices.

Predictions Five: PM servers will imbue IP-PBXs with the ability to intelligently route messages to the appropriate person, at the appropriate time, to the appropriate device.

14.2.6 What will it be like in 2010?

All your 'Net devices will have a PM client you log into. You will communicate with the communications server via your PM client, setting you communication status throughout the day. Also, to present a different "face" on the network, you will have a collection of Avatars, such as "Joe Bloggs—Business" or "Joe Bloggs—Home." The communications server will decide, based on the privacy settings of the Avatars you have created via the PM client, how to contact you. For example, the business Avatar allows you to receive calls on a company extension and all you messages originate as if from the business.

Presence Management will also facilitate communication directories. The Avatar will be a single contact point and you will manage, via the PM client, all your telephone numbers, IM IDs, e-mail addresses, fax numbers, and so on. Communication directories will be lists of Avatars and people can search them, much like one can search ICQ. And like ICQ, different Avatars will be searchable or not. Place Avatars on your temporary "buddy-list" and not only can you connect to the person, but you can also monitor in real-time whether the person is available for communications.

While the PM aspect of the communications server will add presence management, the communications server will also be a dynamic message broker. Messages will be routed appropriately and converted to proper formats, if necessary. For example, an Instant Message can be converted to speech, or a voice message can be converted to text, depending on your contact setting and device currently available.

14.2.7 What is it going to take?

The steps needed to reach this new level of communications are:

- Communication servers will need to do more than just manage and route calls, faxes, and voice mail.
- Communication servers will need to make connections to the data world and converge all messaging. VoIP is the way.
- Service providers will have to agree on PM and IP-PBX interoperability standards.
- IP-enabled devices need to be able to provide PM client software to communicate with the communications server.
- Finally, directory listings need to be accessible and managed by the subscriber (privacy, Avatars).

14.2.8 Summary of predictions

The cost savings and flexibility of an all IP voice systems will be the foundation for the next generation of telephony. The IP-PBX is here and over the next 5 years will become the prevalent, if not the dominant method for businesses to serve telephony applications. Ten years from now, a single pipe will offer commercial and consumer telephony. ISPs, cable companies, and telcos will provide enhanced IP-based communication services via landline and wireless IP connections. In 15 years, IP-based communication servers will be at the heart of ALL communications and the global IP network will provide the connections. IP-based communication servers will enhance, improve, and tie together the deployment of network-based enhanced services. PM servers will imbue IP-PBXs with the ability to intelligently route messages to the appropriate person, at the appropriate time, to the appropriate device—anywhere. I can't wait.

14.3 Brian McConnell's Vision

14.3.1 The Broadband Wireless Future

What does the future hold for telephone systems in the next 5 to 10 years? It's safe to say that we are in the midst of an unprecedented technological revolution in the computing and telecommunications industry. The fruits of this revolution for the consumer are many. As we've seen in the computer industry, prices have come down dramatically while the capabilities of computing technology have increased exponentially. We're now at the beginning of a similar phase in the telecommunications industry.

Probably the most important thing happening in telecom is the widespread deployment of cheap, broadband data networks based on technologies like DSL (digital subscriber line), high speed wireless data links and cable modems. Within a few years, most users in North America will be able to order a 1 megabit to 6 megabit Internet connection for a cost of $200 per month (perhaps less).

The development of high speed wireless communication systems is another important trend that will influence the development of phone systems in the coming years.

Compare this to the recent era when the fastest affordable Internet connections ran at speeds that topped out at about 100,000 bits per second. These new technologies represent a 10 fold to 100-fold improvement in speed, while prices will remain constant. These technologies, as fast as they are, will later be replaced by even faster services such as VDSL and fiber to the curb (with the ability to deliver data at speeds from 10 megabits to 1 gigabit, yet another 10 fold to 100 fold improvement).

What does this era of cheap, over-abundant bandwidth mean for the buyer? It means that video, not voice, will become a common means of communicating in a work environment. Videoconferencing will not replace the telephone, but combined with desktop computers, it will compliment conventional voice telephony, and will become a widely used tool. As anyone who has used a good videoconferencing system will attest, it's a much better way to participate in group discussions (compared to being a disembodied voice coming out of a Polycom speakerphone/flying saucer device).

High quality videoconferencing, with picture quality comparable to a VHS videotape, requires about 400,000 bits per second for each session. Until recently, this was out of reach for most small businesses because the equipment and data connections were so expensive. Today, set-top videoconferencing kits can be bought from vendors like Polycom for about $600 (Polycom is releasing a PC videoconferencing kit in Fall 2000). High speed digital subscriber line connections can easily carry the amount of data required for video calls. So, even today, high quality videoconferencing is accessible for even the small business.

Over the next 2 to 3 years, the cost and complexity of videoconferencing terminals will continue to decrease, and will reach a point where the added cost of equipping networks with video will be an incremental expense. The cost of high speed networks and interconnections to the public Internet will also continue to decrease as new technologies and competition upset the once air-tight Bell Company monopolies. As this happens, video will become a standard feature in business telephone systems.

If you pop the cover off these future telephone systems, you'll find that, like virtually every other computing or communication device, they will be using Internet technology (e.g. Ethernet combined with the TCP/IP protocol) to transport phone calls, video calls, and the information associated with them. You will no longer have separate wiring for your phones and computers. You'll simply provide each user with an Ethernet port, to which they will be able to connect their computer, telephone, video terminal, printer, and so forth.

Much work has already been done to develop standards for transporting voice and video phone calls over TCP/IP networks (either private LANs or the Internet). These standards, among them H.323, SIP and MGCP, are currently being widely used. They will continue to mature, and will soon be able to do pretty much everything that a digital PBX phone can do. As these standards mature, you'll see more and more vendors migrating away from proprietary digital telephones that require their own cabling to IP telephones and videophones that connect directly to a company's data network.

These systems will combine telephones, videophones and telecom- enabled PCs into a single integrated intercom system. So, if a user with a videophone calls a user with an ordinary phone, they'll be able to talk to (but not see) each other. Or, if a videophone user calls other videophone users, they'll automatically be connected in a voice + video call.

As prices decline and speeds increase, we'll reach a point within the next five years where videoconferencing will escape from the boardroom to become a standard desktop option throughout most companies.

Wireless Communication

Another important trend in telecom is the development of high speed wireless communication systems (both public and private). This technology will enable phone system vendors to provide users with mobility that was, until recently, almost unheard of except in the most expensive systems.

Several wireless technologies will take up an important role in the workplace: Bluetooth, DECT, site-based cellular networks and wireless Ethernet. None of these technologies is, by itself, a magic bullet that solves every communication issue single-handedly. Each has its own application or sweet spot. All of them will become common options in most business phone systems.

Bluetooth, developed by Ericsson and a consortium of telecom and consumer electronics companies, is a short-range wireless data technology. It is designed to provide a relatively high speed wireless connection (up to about 1 Megabit) between personal communication devices, up to a range of about 100 feet (though it typically operates over much shorter ranges). Bluetooth, for example, will be widely used for automatic data synchronization (e.g. you sit down at your desk and your PDA quietly talks to your PC to sync its contact list and calendar with your Microsoft Outlook contacts and schedule). Bluetooth is not primarily intended to transport voice calls. However, it will be widely used in cellular phones, PDAs, and other portable devices to make it easy to synchronize data.

DECT, a European standard for multi-user cordless telephones, is used to build a sort of poor man's cellular network in an office environment. This system allows many cordless phones to share one or more base stations, and also allows phones to roam from one cell to another (just as they do in a cellular network). This enables phone system vendors to offer digital cordless handsets as an option, and to allow users to roam throughout a facility and always be available to receive calls. Although the DECT standard has not been widely adopted in the US, systems derived from it (such as the Siemens Gigaset product line) have been quite successful. Expect this technology to become more sophisticated and for cordless handsets or combination corded-cordless handsets to become a standard option for future phone systems.

Site based cellular networks take the concept behind DECT and extend it even further. They allow users to take their office phone with them when they leave the building, and to automatically roam onto the public cellular network. This is an especially useful feature for companies where the users are highly mobile, and are frequently off-site (though it is often easier to set a conventional phone system up to forward users calls to their offsite number when they are out of the office). Expect to see more multi-mode cordless phones that can operate on both public and private wireless networks in the upcoming years.

Wireless Ethernet is another important technology that is being adopted by phone system vendors. Wireless Ethernet, though originally intended for wireless data applications, can also transport IP-based voice calls. One company, Symbol Technologies, has already created a line of wireless Ethernet phones that work with several next generation phone systems. One of the neat things about wireless Ethernet is its seamless interoperability with conventional wired Ethernet and TCP/IP networks. Any phone system that is capable of handing calls off to IP-based telephones will be able to send calls to wireless Ethernet phones (the fact that the underlying connection is wireless is completely transparent to the phone system). Ericsson has already incorporated support for wireless Ethernet into its breakthrough WebSwitch phone system.

Plug and Play Systems

Another trend that will influence the design of phone systems is the concept of modularity. Most data networking products have become very easy to install and maintain in recent years (Ethernet hubs and print servers are a good example of this). Telephone system vendors are also beginning to realize that, if they make their systems easier to install and configure, their systems will be more appealing to end users, especially budget conscious small businesses.

Systems such as the Shoreline Communications' solution are excellent examples of this trend. This system is built to work like datacom components such as Ethernet hubs. Instead of connecting every single user to a single, mainframe-like box, the system is based on a distributed design where 8 to 12 users are connected to pizza box size appliances which are, in turn, connected to a company's Ethernet backbone. Many of these systems can be interconnected via the data network to form a single unified system.

This design approach has the benefit of increasing reliability (if one unit fails, the others continue to function), and also makes installation a lot easier. It simplifies wiring, and makes it easy for companies to relocate users quite easily. This type of design will become a *de facto* standard for small and medium size systems in the next few years.

14.3.2 What to Look For

If you're shopping for a system for your company, there are a few things you definitely want to look for. Few systems on the market today provide all of the features described so far. What is important is to select a system that is capable of evolving to support new technologies as they emerge. Most traditional phone systems lock you into a closed hardware design that cannot easily be upgraded to support new technologies such as voice over IP. Newer systems are much more flexible in this respect.

PC based telephony servers and hub-like networking devices are typically designed so that both the hardware and software can be upgraded in the field. Whether a new feature requires a hardware or software upgrade depends on what you're trying to accomplish. Often, the most useful features (such as smart call forwarding) can be implemented entirely in software, and can be added to a system at little or no cost. Other features, such as IP telephony and videoconferencing, require specialized hardware, and will often require a hardware upgrade to the system.

How do you figure out if a system is software upgradable? A good way to find this out is to talk to customers who are using the system. Every vendor will lie to you when they answer this question. Even legacy PBXs are upgradable, but you need specialized training to learn how to install software upgrades and to program the systems. What you want to know is whether the system can be upgraded easily (i.e. by downloading an update kit from the vendor's Web site and running an installation script). Some systems make this fairly easy to do. With others, it can be a real nightmare. The only way to really figure this out is to talk to people who actually deal with the system on a daily basis.

In the future, vendors will make the upgrade process automatic. The phone system will periodically talk to the "mothership" (a server at the vendor's location), and automatically download available upgrades. In fact, one vendor, BizFon, is already heading in this direction by remotely administering its customer's systems. This capability will become more common, and more sophisticated, in the upcoming years.

The ability to upgrade hardware in the field is also a very important selection criteria. This is where a modular system design really comes in handy. Take Shoreline Communications' system as an example. Their basic building block is a 8-12 user pizza box appliance that connects to a company's data network. Let's say that two years from now, their customers are clamoring for multi-line cordless phones that use a standard similar to DECT. The vendor would be able to add support for this hypothetical option by creating a new appliance—a base station designed to talk to cordless phones.

Whether the vendor will actually do this is another question. However, if their system is based on connecting many smaller units to a company's data network to form an integrated system, it is easy for them to extend their product's capabilities by creating new types of telecom appliances to connect to this network.

On the other hand, if the vendor consolidates the entire system into a single, mainframe like box, like a PBX, it may be harder for them to provide hardware upgrade paths for users. This is especially true of black-box systems, like key systems, where everything is contained in a single case that cannot be opened by the user. PBXs are a little more upgradeable since they allow the user to mix and match different types of line cards within a chassis.

Support for IP telephony and videoconferencing standards is an absolute must. If a vendor tells you that they don't support voice over IP because it's not as reliable as conventional telephony, they are selling you a bill of goods. What they're really saying is "We haven't bothered to invest in making our system future-ready, but we'll try to scare you away from buying competing products." The key standards to look for are: H.323 (version 2 or later), SIP (session initiation protocol) and MGCP. Of the lot, H.323 is the most widely used standard at the moment, although this will change with time. The important thing is that they support one or more TCP/IP based systems for routing phone and video calls over data networks. If they don't do this, or force you to use their own proprietary format, keep looking.

The most important thing isn't to find a system that is loaded down with every feature imaginable, but to find a system that is based on an open and expandable design. These systems will be able to grow with your business, and will evolve to support new technologies, many of which are just drawings on a whiteboard as you are reading this today.

About the Authors

Dave Krupinski is a software engineer, author, and consultant. Recently, Dave has been developing some of the most popular web sites on the Internet, including DirectHit.com. Previously, Dave was the vice president of marketing at PhoneZone.com (acquired by Hello Direct, Inc.) and the vice president of marketing and director of business development for Artisoft, Inc. in Europe. Dave's background also includes designing and implementing large-scale information systems and computer telephony projects for Fortune 500 companies and other large organizations around the world.

Charlie Schick, Ph.D. is an independent telecom analyst, concentrating on wireless voice and data technologies, ubiquitous computing, and computer telephony. His writings appear in many online and offline publications, such as HelloDirect.com and ThinkMobile.com, and in his TechCheck column in *Computer Telephony Europe*. Charlie also has a participating interest in two startups - TrekMail, a voice to e-mail service, and JackFinder.

Brian McConnell is the founder of PhoneZone.com (now part of Hello Direct, Inc.), and one of the telecom industry's leading experts on emerging technologies. Brian writes about telecom technology for numerous online and print publications, and also teaches telecom-related classes at networking and telecom conferences worldwide.

Appendix A

Acronym Reference

The following is a reference for acronyms used in this book and throughout the telecommunications industry.

ACD - Automated Call Distribution

ADPCM - Adaptive Differential Pulse Code Modulation

API - Application Programming Interface

ATM - Asynchronous Transfer Mode

BRI - Basic Rate Interface

CO - Central Office

COM - Component Object Model

CSTA - Computer Supported Telephony Architecture

CTI - Computer and Telephone Integration

DID - Direct Inward Dial

DPNSS - Digital Private Networking Signaling Systems

DSL - Digital Subscriber Line

ECTF - Enterprise Computer Telephony Forum

GUI - Graphical User Interface

IEFT - Internet Engineering Task Force

IP - Internet Protocol

ISDN - Integrated Services Digital Network

ISP - Internet Service Provider

ITU - International Telecommunications Union

IVR - Interactive Voice Response

JTAPI - Java Telephony Application Programming Interface

LAN - Local Area Network

LDAP - Lightweight Directory Access Protocol

MAPI - Messaging Application Programming Interface

MVIP - Multi-Vendor Integration Protocol

PBX - Private Branch eXchange

PC - Personal Computer

PCI - Peripheral Component Interface

PCM - Pulse Code Modulation

PIM - Personal Information Manager

POTS - Plain Old Telephone Service

PRI - Primary Rate Interface

PSTN - Public Switched Telephone Network

RAID - Redundant Array of Inexpensive Disks

SCSA - System Computing Signaling Architecture

SCTP - Simple Computer Telephony Protocol

SME - Small and Medium Enterprise

TAPI - Telephony Application Programming Interface

TDM - Time-Division Multiplexing

TSAPI - Telephony Server Application Programming Interface

VoDSL - Voice over Digital Subscriber Line

VoIP - Voice over Internet Protocol

WAN - Wide Area Network

Appendix B

Additional Resources

For additional information about the topics discussed in this book, you are encouraged to visit the following web sites:

3Com Corporation - http://www.3com.com

Aculab, plc. - http://www.aculab.com

Alcatel - http://www.alcatel.com

AltiGen Communications, Inc. - http://www.altigen.com

Artisoft, Inc. - http://www.artisoft.com

C-Phone Corporation - http://www.cphone.com

Cisco Systems, Inc. - http://www.cisco.com

COM2001.com - http://www.com2001.com

Comdial Corporation - http://www.comdial.com

Connectix Corporation - http://www.connectix.com

CosmoCom, Inc. - http://www.cosmocom.com

dba Telecom Corporation - http://www.dba-tele.com

Dialogic Corporation - http://www.dialogic.com

Ericsson - http://www.ericsson.com

Hello Direct, Inc. - http://www.hellodirect.com

Interactive Intelligence, Inc. - http://www.inter-intelli.com

InVADE International - http://www.invade.net

Jetstream Communications - http://www.jetstream.com

Linx Communications, Inc. - http://www.linxcom.com

Lucent Technologies - http://www.lucent.com

Lucent Technologies Alchemy Series - http://www.networkalchemy.co.uk

Merlot Communications - http://www.merlot.net

Microsoft Corporation - http://www.microsoft.com

Mitel Corporation - http://www.mitel.com

Motion Media Technology, Ltd. - http://www.motionmedia.co.uk

Natural Microsystems Corporation - http://www.nmss.com

Netergy Networks, Inc. - http://www.netergynetworks.com

Nexus Telecom, Ltd. - http://www.nexustelecomltd.com

Nortel Networks - http://www.nortel.com

PIKA Technologies, Inc. - http://www.pika.ca

Polycom, Inc. - http://www.polycom.com

Praxon, Inc. - http://www.praxon.com

Quicknet Technologies, Inc. - http://www.quicknet.net

RADVision - http://www.radvision.com

RING! - http://www.ringrosa.com

Shoreline Communications, Inc. - http://www.shoretel.com

Siemens - http://www.siemens.com

Telogy Networks - http://www.telogy.com

Tennyson Networks, Pty. Ltd. - http://www.tennyson.com.au

ThinkLink, Inc. - http://www.thinklink.com

Vertical Networks, Inc. - http://www.verticalnetworks.com

VirtualPBX.com, Inc. - http://www.virtualpbx.com

VocalTec Communications, Ltd. - http://www.ocaltec.com

VSN Systemen BV - http://www.opentsp.com

Webley Systems, Inc. - http://www.webley.com

Wildfile Communications, Inc. - http://www.wildfire.com

INDEX

IP Based Services:

VPN

Voice

Unified Messg

IP PBX

IP CTX

Voice over managed IP networks

- Web Call center